PHYSICS RESEARCH AND TECHNOLOGY

IONIZING RADIATION

ADVANCES IN RESEARCH AND APPLICATIONS

PHYSICS RESEARCH AND TECHNOLOGY

Additional books in this series can be found on Nova's website under the Series tab.

Additional e-books in this series can be found on Nova's website under the eBooks tab.

PHYSICS RESEARCH AND TECHNOLOGY

IONIZING RADIATION

ADVANCES IN RESEARCH AND APPLICATIONS

TAMAR REEVE
EDITOR

nova
science publishers
New York

Library of Congress Cataloging-in-Publication Data

ISBN: 978-1-53613-539-8
ISSN: 2473-3679

Published by Nova Science Publishers, Inc. † New York

CONTENTS

PREFACE

In this compilation, the authors examine the importance of ionizing radiations for thermoluminescence dosimetry that is the current area of research for medical and industrial purposes. Ionizing radiations are harmful to the human body, so, there is a need to measure small doses in the environment as well as very high doses at the time of accident like radiation leakage and for the treatment of cancer. Next, recent advances regarding effects on exposure of some foods to ionizing radiations are presented. The dosage required for complete sterilization may, at times, lead to undesirable changes in food flavours or may exceed the permitted levels. Combining irradiation with other treatments yields satisfactory results in these cases. Combined applications of ionizing radiation with heat, low temperature, high hydrostatic pressure, and modified atmospheres are also discussed. Later, new results concerning cell survival and genetic instability of wild-type and radiosensitive yeast cells of *Saccharomyces cerevisiae* surviving after irradiation with 60Co γ-rays, 239Pu α-particles and 254 nm UV light are presented. Survival was determined by cell ability to produce macrocolonies on a solid nutrient medium. The authors also review data on the radiation resistance of AlGaN/GaN and InAlN/GaN High Electron Mobility Transistors (HEMTs) as well as emerging Ga2O3 photodetectors and rectifiers to different types of ionizing radiation. Both of these wide bandgap semiconductor (nitride-based and gallium oxide-based) materials are much more radiation-hard than GaAs or Si and this is largely a result of their high bond strengths. By using the test system of blood lymphocytes of healthy individuals, the following chapter presents the co-mutagenic (potentiating) effect of the drugs – a calcium channel blocker verapamil and an antioxidant ascorbic acid-on the radiosensitivity of cells. It was found that when the lymphocytes are irradiated in a small dose (0.3 Gy), ascorbic acid (80.0 μg/blood) and verapamil (4.0 μg/ml blood) increase the frequency of chromosomal aberrations in comparison with the radiation effect by 75 and 62, 5% respectively. Afterwards, the experimental data of the effects of temperature, light intensity, background ionizing and non-ionizing radiation on physicochemical properties of water, water solution and cell function are presented. Considering the fact that the cell membrane permeability for water is more than 10 times higher than for inorganic ions, it is hypothesized that, upon the effect of chemical and physical factors, the changes of net water efflux through the membrane precede the activation of ionic fluxes in the membrane. Also in this compilation, a study is included which investigated the modifying effect of astaxanthin on radiation induced genome damages in cultivating human peripheral blood lymphocytes (PBL). The study indicated that astaxanthin in concentration of 20.0 μg/ml demonstrated evident radioprotective properties by reduction of the ChA level, decreasing of DNA damages and increasing of the apoptotic rate. In the last chapter, the authors present some works that study the causes of the high resistance of IRRB to ionizing radiation. Then the authors focus on presenting in silico approaches that use protein sequences of bacteria in order to predict if an unknown bacterium belongs to IRRB or ionizing-radiationsensitive bacteria (IRSB).

Chapter 1 - Ionizing radiation is generally considered to be the form of radiation exposure that will produce a subatomic ionization event in the target molecule. Ionizing radiations are harmful to the human body, so, there is a need to measure small doses in the environment as well as very high doses at the time of accident like radiation leakage and for the treatment of cancer. Thermoluminescence (TL) is one of the important technique to estimate the quantity of absorbed dose. Today, there are number of commercial TL dosimeters readily available for users as TLD badges and are mostly based on oxides, fluorides and sulphates: LiF: Mg,

Cu, P (TLD-700H), Al_2O_3 (TLD- 500), $CaSO_4$:Dy (TLD-900) and CaF_2:Dy (TLD-200) etc. Continuous efforts have been made by the research community worldwide to develop new materials and to improve dosimetric properties of already available materials as an efficient TLD material that can be used in different areas with low or high levels of radiations such as photons, protons, swift heavy ions etc. The present chapter deals with the study of importance of ionizing radiations for thermoluminescence dosimetry that is the current area of research for medical and industrial purposes.

Chapter 2 - Research has established food irradiation as a technically feasible option in the food industries. Irradiation is a cold process that improves the safety and quality of food commodities. Ionizing radiations, such as gamma ray, electron beam and X-rays have been approved for decontaminating packaged/frozen foods, feeds, and packaging materials. These methods are particularly effective in destroying harmful organisms in raw and processed materials. Approval of irradiation treatments in the food sector is linked with regulation-compliant labelling, availability of detection methods, and international trade regulations. Recent advances regarding effects on exposure of some foods to ionizing radiations are presented. The dosage required for complete sterilization may, at times, lead to undesirable changes in food flavours or may exceed the permitted levels. Combining irradiation with other treatments yields satisfactory results in these cases. Combined applications of ionizing radiation with heat, low temperature, high hydrostatic pressure, and modified atmospheres are also discussed. Recently, safety of packaging systems has also been taken into consideration to guarantee the safety of using radiation for sterilization of food and healthcare products. Irradiation of pre-packaged foodstuffs may generate radiolytic by-products in the packaging material, which may compromise their barrier properties and allow the radiolytic products to enter the foodstuffs. Technical challenges regarding the safety of food packaging materials have also been elaborated. Moreover, innovative electronic sensing techniques, such as electronic nose and electronic tongue, for authentication and control of irradiated foods are discussed for potential future applications.

Chapter 3 - New results concerning cell survival and genetic instability of wild-type and radiosensitive yeast cells of *Saccharomyces cerevisiae* surviving after irradiation with ^{60}Co γ-rays, ^{239}Pu α-particles and 254 nm UV light are presented. Survival was determined by cell ability to produce macrocolonies on a solid nutrient medium. The diploid strain exhibited sigmoid shape of the dose-effect curve independently of radiation type. The haploid strain showed exponential survival curve both after sparsely and densely ionizing radiations while the shape of survival curves was sigmoid after UV exposure. Genetic instability was defined by the delayed appearance of clones by cells surviving irradiation. The effect of delayed colony appearance by irradiated cells was revealed to be more expressed for the diploid wild-type cells characterizing by a sigmoid shape of the dose-effect curve and capable of the diploid-specific recovery in contrast to isogenic haploid strain (100% vs. 20-30%) incapable of such kind of recovery and characterizing by the exponential survival curve. Similar experiments were performed for haploid and homozygous diploid radiosensitive mutants showing the exponential dose-effect curves. On the contrary, UV sensitive mutant exhibits the sigmoid dose-effect curve like a diploid strain. Here again the degree of genetic instability was mainly determined by cell ploidy since both resistant and radiosensitive diploid strains in contrast to haploids demonstrate large extent of the delayed appearance of clones by cells surviving after irradiation with UV light (100% vs. 30%). This means that, in spite of the defects in the repair of DNA double strand breaks, the irradiated cells of the diploid but not haploid strains inherit in a greater extent of some sub-lesions responsible for genetic instability.

Relative biological effectiveness (RBE) of α-particles defined by the ratio of isoeffective dose of γ-rays и α-particles was shown to be of 4.6 ± 0.2 for wild-type diploid cells and only 2.0 ± 0.2 for all haploid strains and radiosensitive diploid strains. It indicates that RBE values are greatly determined by cell ability to recover from radiation damage. These findings are not new for cell survival, while they are fundamentally new for genetic instability.

The dependence of the delayed colony appearance of diploid and haploid yeast cells on their survival was substantially the same after exposure to UV-light, sparsely and densely ionizing radiation. Since at the same survival after exposures to radiations of various quality equally effective number of lethal damage was induced, these data indicate that the identical number of the accompanying sub-lesions responsible for the delayed colony appearance are produced in irradiated cells.

The final conclusion is that genetic instability is mainly determined by cell ploidy rather than the shape of survival curve and the ability of cell to recover from radiation damage as it is traditionally assumed for *Saccharomyces cerevisiae* yeast cells.

Chapter 4 - The authors review data on the radiation resistance of AlGaN/GaN and InAlN/GaN High Electron Mobility Transistors (HEMTs) as well as emerging Ga_2O_3 photodetectors and rectifiers to different types of ionizing radiation. Both of these wide bandgap semiconductor (nitride-based and gallium oxide-based) materials are much more radiation-hard than GaAs or Si and this is largely a result of their high bond strengths. The states in the bandgap introduced by ionizing or non-ionizing radiation and comparative figures-of-merit such as carrier removal rates in these materials are covered. The carrier removal rates in GaN, for example, are a function of initial carrier concentration and dose but not of dose rate or hydrogen concentration. Device simulators are able to reproduce the observed changes in HEMTs after radiation-damage. Non-ionizing radiation like neutrons leads to so-called Gossick disorder zones of extended damage and at high doses can pin the Fermi level. By contrast, ^{60}Co γ-rays produce less reduction in HEMT drain current than other types of radiation. Optical devices such as InGaN/GaN blue Light-Emitting Diodes (LEDs) irradiated with either protons (fluences near 10^{14} cm^{-2}) or electrons (fluences near 10^{16} cm^{-2}) exhibit degraded current-voltage and light output-current characteristics with increasing dose. A comparison of optical and electrical performance of the LEDs shows the former is more sensitive to change than the latter with both proton and electron irradiation. Ga_2O_3 shows a similar level of radiation resistance to GaN. The measured carrier removal rate for 10 MeV electrons in this material was 4.9 cm^{-1}.

Chapter 5 - By using the test system of blood lymphocytes of healthy individuals, this chapter presents the co-mutagenic (potentiating) effect of the drugs – a calcium channel blocker verapamil and an antioxidant ascorbic acid-on the radiosensitivity of cells. It was found that when the lymphocytes are irradiated in a small dose (0.3 Gy), ascorbic acid (80.0 μg/blood) and verapamil (4.0 μg/ml blood) increase the frequency of chromosomal aberrations in comparison with the radiation effect by 75 and 62, 5% respectively. With an augmentation of the radiation dose, these drugs (depending on their concentration) also increase the frequency of chromosome aberrations. With a raise in radiosensitivity of lymphocytes (cell irradiation in the G_2-period of the cell cycle), the co-mutagenic effects of verapamil and ascorbic acid are more pronounced, as compared with the radiation in the G_0-period. This is the first time that the dependence of the formation of co-mutagenic effects in irradiated human cells was established: a dependence on the dose of ionizing radiation, the degree of radiosensitivity of cells, the concentration of drugs with co-mutagenic activity and the specificity of their action. This can be considered as an experimental justification for the correction of the designation of the drugs with identified co-mutagenic properties for the people that work in the sphere of action of ionizing radiation and the residents of territories contaminated with radionuclides due to the Chernobyl disaster.

Chapter 6 - The Membrane Theory, which is based on the principles of classical thermodynamics, explains the biological effects of environmental factors on living cells by the changes of membrane electrical conductivity. Thus, the experimental data on biological effects of extremely weak physical signals (with the intensity even less than thermal threshold), which are unable to activate the ionic channels in membrane and have non-linear dose-dependent effect on cell, cannot be explained from the point of this theory. Therefore, the messenger(s), through which the biological effect of such weak signals on cells and organisms is realized, should have quantum-mechanical character.

It is known that the valence angle of a water molecule, determining its physicochemical properties, is extremely sensitive to weak chemical and physical factors. At present, it is also well established that cell hydration is a fundamental parameter determining cell functional activity, which is realized by regulating the activity of intracellular macromolecules by their folding-unfolding mechanism as well as by the membrane surface-dependent changes of the number of functionally active protein molecules in membrane, having enzymes, receptors and ionic channels-forming properties. Considering the fact that the cell membrane permeability for water is more than 10 times higher than for inorganic ions, it is hypothesized that, upon the effect of chemical and physical factors, the changes of net water fluxes through the membrane precede the activation of ionic fluxes in the membrane.

For supporting this hypothesis, in this chapter the experimental data of the effects of temperature, light intensity, background ionizing and non-ionizing radiation on physicochemical properties of water, water solution and cell function are presented.

Chapter 7 - *Purpose:* Astaxanthin (a carotenoid of xanthophyll group) satisfies all requirements demanded to potential radioprotectors. The purpose of the authors' study was to investigate a modifying effect of astaxanthin on radiation induced genome damages in cultivating human peripheral blood lymphocytes (PBL).

Materials and methods: To establish astaxanthin radioprotective activity, the authors used method for scoring of structural chromosomal aberrations (ChA) in PBL and neutral single-cell gel electrophoresis (comet assay) for estimation of the level of DNA breaks in cells (expressed as tail moment (TM) for "classical comets" and rate of "atypical comets" (AC) for cells with greatly fragmented DNA). Blood samples were obtained by venepuncture from five conditionally healthy volunteers. Astaxanthin in final concentrations of 2.0, 10.0, 20.0 and 40.0 μg/ml had been added into the culture medium before cultivation began, prior to γ-irradiation of PBL (dose 1.0 Gy). Cultures were incubated at 37°C for 48-50 h. Chromosome slides were prepared after 50 h of cultivation. Comet assay was carried out before cultivation and on 48 h of culturing.

Results: Astaxanthin in all studied concentrations *per se* had no mutagenic activity and obviously reduced the level of radiation induced chromosomal aberrations. The most efficient concentration of astaxanthin was 20.0 μg/ml. The effect of astaxanthin in this concentration on exposed cells manifested in significant decreasing of the radiation-induced ChA level: from 26.06 ± 1.81 per 100 metaphases (phm) to 9.03 ± 0.73 phm (p<0,001). Comparative analysis of the ChA spectra indicated the most substantial reduction of classic unstable cytogenetic markers of radiation exposure (dicentric and ring chromosomes). Likewise, statistically significant decrease in TM in irradiated lymphocytes treated by astaxanthin compared with untreated irradiated samples was observed both before cultivation (3.21 ± 0.48 and 6.55 ± 1.82 respectively, p<0,05) and after 48 h of cultivation (5.27 ± 1.77 and 12.86 ± 0.74 respectively, p<0,05). The level of AC, which evidently arose from apoptotic cells, in irradiated cultures treated by astaxanthin was approximately two times higher as compared with irradiated samples (7.15 ± 1.13% and 3.57 ± 0,81% respectively, p<0,05).

Conclusions: Our study indicated that astaxanthin in concentration of 20.0 μg/ml demonstrated evident radioprotective properties by reducing the level of ChA, decreasing DNA damage and increasing the apoptotic rate.

Chapter 8 - Ionizing-radiation-resistant bacteria (IRRB) could be used for bioremediation of radioactive wastes and in the therapeutic industry. Limited computational works are available for the prediction of bacterial ionizing radiation resistance (IRR). In this chapter, the authors present some works that study the causes of the high resistance of IRRB to ionizing radiation. Then the authors focus on presenting in silico approaches that use protein sequences of bacteria in order to predict if an unknown bacterium belongs to IRRB or ionizing-radiation-sensitive bacteria (IRSB). These approaches formulate the problem of predicting bacterial IRR as a multiple instance learning (MIL) problem where bacteria represent the bags and primary structure of basal DNA repair proteins of each bacterium represent the instances inside the bags. The authors also present a formulation of the problem of MIL in sequence data and explain how it could be used to solve the problem of IRR prediction in bacteria. A brief comparison of the presented approaches is provided.

In: Ionizing Radiation
Editor: Tamar Reeve
ISBN: 978-1-53613-539-8

Chapter 1

THE IMPORTANCE OF IONIZING RADIATIONS FOR THERMOLUMINESCENCE DOSIMETRY

***Vibha Chopra*[1] *and S. J. Dhoble*[2,*]**

[1]P.G. Department of Physics & Electronics,
DAV College, Amritsar, Punjab, India
[2]Department of Physics, R.T.M.
Nagpur University, Nagpur, India

ABSTRACT

Ionizing radiation is generally considered to be the form of radiation exposure that will produce a subatomic ionization event in the target molecule. Ionizing radiations are harmful to the human body, so, there is a need to measure small doses in the environment as well as very high doses at the time of accident like radiation leakage and for the treatment of cancer. Thermoluminescence (TL) is one of the important technique to estimate the quantity of absorbed dose. Today, there are number of commercial TL dosimeters readily available for users as TLD badges and

* Corresponding Author Email: sjdhoble@gmail.com.

are mostly based on oxides, fluorides and sulphates: LiF: Mg, Cu, P (TLD-700H), Al_2O_3 (TLD- 500), $CaSO_4$:Dy (TLD-900) and CaF_2:Dy (TLD-200) etc. Continuous efforts have been made by the research community worldwide to develop new materials and to improve dosimetric properties of already available materials as an efficient TLD material that can be used in different areas with low or high levels of radiations such as photons, protons, swift heavy ions etc. The present chapter deals with the study of importance of ionizing radiations for thermoluminescence dosimetry that is the current area of research for medical and industrial purposes.

Keywords: ionizing radiation, thermoluminescence, radiation dosimetry, LiF

1. Introduction

Radiation is a form of energy that comes from a source and may penetrate various materials. Generally radiations are categorized as (a) Charged particulate radiation that includes fast electrons, heavy charged particles (b) Uncharged radiations that includes electromagnetic radiations and neutrons. Further fast electrons are known as beta particles (+ve or –ve) emitted in nuclear decay. Heavy charged particles denotes a category of ions having mass one atomic mass unit or greater such as alpha particles, protons, fission products etc. On the other hand an uncharged radiation includes X-rays (that are emitted in the rearrangement of electron shells of atoms) and gamma rays (that originate from transitions within the nucleus).

Generally radiation is also categorized as either ionizing or non ionizing depending upon the energy. The energy range of interest, ranges from 10 eV to 20 MeV. The lower energy bound is set by the minimum energy required to produce ionization in the materials by radiation that interacts with it. So, the radiation with energy greater than this minimum value are categorized as Ionizing Radiations.

1.1. Ionizing Radiations

Ionizing radiations carries more than 10 eV energy that is enough to break the chemical bonds and knock out electrons from the atoms and create ions. On the other hand non ionizing radiations are the one for which kinetic energy of particles is too small to produce ions when passing through matter. They have only sufficient energy to change the rotational, vibrational or electronic valence configurations of atoms. Non ionizing radiation is capable of causing thermal ionization, if it deposits enough heat to raise the temperatures to ionization radiation energies, but this is possible at very high energies as compare to ionizing radiation that requires only single particle to cause ionization.

From electromagnetic spectrum, the radio waves, microwaves, infrared rays and sometimes visible light comes under the category of non ionizing radiations. The lower frequencies of UV light may cause chemical changes and molecular damage similar to ionization but still it is technically considered as non ionizing. Further the high frequency UV rays, X rays and gamma rays are ionizing radiations. Ionizing radiations have great potential of biological damage and hence cause risk of cancer. High doses of ionizing radiation exposure can result in acute radiation syndrome (ARS) with skin burns, hair loss, internal organ failure and death. Ionizing radiations are harmful to the human body, so, there is a need to measure even small doses in the environment and very high doses at the time of accident like radiation leakage and for the treatment of cancer. Thermoluminescence (TL) is one of the important technique to estimate the quantity of absorbed dose. Today, there are number of commercial TL dosimeters readily available for users as TLD badges. Inspite of having high risk, radiations are used for diagnosis, treatment and research. Ionizing radiations are used for cancer treatment, to determine the composition of materials in the process called neutron activation analysis, to study the age of old materials by radiocarbon dating etc. Since radiation is very useful topic to study so it is important to be familiar with the fundamental mechanisms by which radiation interacts and lose their energy in matter. This chapter includes the study of basic interactions of ionizing radiations with matter.

1.2. Interaction of Radiation with Matter

When the ionizing radiation encounters matter, the effects are produced by different processes. These processes are the basis of all the particle detectors. As explained earlier there are two major categories of radiations, charged particulate radiation and uncharged radiations. In case of charged particulate radiation, an electric charge carried by the particle continuously interacts with the electrons (through Coulomb force) present in any medium through which they pass. In general two principal features form the basis of interaction of charged particle with matter (1) loss of energy by the particle (2) deflection of particle from its path. These effects are the results of two processes: (a) Inelastic collisions with the atomic electrons of material (b) elastic scattering from nuclei. Other processes are emission of Cherenkov radiation, nuclear reactions and bremsstrauhlung. On the other hand, the uncharged radiations are not subjected to coulomb force. Instead these radiations must first undergo a catastrophic interaction which radically alters the properties of an incident radiation in a single counter. In all the cases, practically the interaction results in either full or partial transfer of energy of an incident radiation to the electrons or nuclei of the target atoms. But sometimes in case of uncharged particles like neutron, gamma ray, the radiations does not interact and hence can pass without an effect through the detector volume. Otherwise, in case of interaction, X or gamma ray can transfer all or part of its energy to an electrons of the target resulting in the secondary electrons that are similar to beta particles whereas the neutrons after interaction, produces secondary heavy charged particles which then serve as base for detector signal (Knoll G.F., 1999; Leo W.R., 1994). So, based on the above introduction, an interaction of heavy charged particles, electrons, photons, neutrons is explained in details.

1.2.1. Interaction of Heavy Ions with Matter

Heavy charged particles such as alpha particles, protons interact with matter, primarily through inelastic collisions. Upon entering the absorbing material, the charge particle interact simultaneously with many electrons resulting into either raising the electron to higher shells with the absorber atom called excitation or completely removing the electron from the atom known as ionization. Energy is transferred from charged particle to electron and hence it results in the decrease in velocity of charged particle. The maximum energy that can be transferred is about 1/500 of the particle energy per nucleon. Since the particle is interacting with many electrons, so the net effect is to continuously decrease its velocity until the particle is stopped. So, the distance beyond which no particle will penetrate is defined as Range.

Now the atomic collisions of charged particle with target are divided into two groups: Soft collision, in which only an excitation occurs and hard collision, in which the energy transferred is sufficient to cause ionization. In some cases, an electron may undergo a large enough impulse so that after leaving its parent atom, it still may have sufficient kinetic energy to create further ions by same mechanisms. These energetic electrons are called Delta rays or Knock on electrons.

Now, inelastic collisions are statistical in nature that occurs with quantum mechanical probability. Since the number of collisions per macroscopic path length is generally large, so, the fluctuations in the total energy loss are small, therefore, the average energy loss per unit path length should be considered and this quantity is known as the stopping power or simply dE/dx. It was first calculated by Bohr using classical mechanics and later corrected by Bethe, Bloch and others using quantum mechanics (Knoll G.F., 1999; Leo W.R., 1994).

The Bohr formula for stopping power or specific energy loss was given as

$$-\frac{dE}{dx} = \frac{Z^2 e^4 n}{4\pi\varepsilon_0^2 m_e v^2} \log_e\left(\frac{2m_e v^2}{h\vartheta}\right) \tag{1}$$

where +Ze is the charge of incident particle, v is the velocity with which it moves, m_e is mass of electron. ϑ is frequency of target electrons with which they revolve around the nucleus of parent atom since they are free, n is the number of electrons per unit volume of the medium.

In deriving this expression, it was assumed that v << c so that $m_e = m_o$ (rest mass of electron). But H. A. Bethe explained that if the velocity of incident charged particle is quite high, a relativistic effect has to be applied. So the relation will become

$$-\frac{dE}{dx} = \frac{Z^2e^4n}{4\pi\varepsilon_0^2 m_e v^2}\log_e\left(\frac{2m_e v^2}{h\vartheta}\right) - \log_e\left(1-\frac{v^2}{c^2}\right) \quad (2)$$

However the exact quantum mechanical relation was developed by F. Bloch. The exact relation is

$$-\frac{dE}{dx} = \frac{Z^2e^4n}{4\pi\varepsilon_0^2 m_e v^2}\log_e\left(\frac{2m_e v^2}{h\vartheta}\right) - \log_e\left(1-\frac{v^2}{c^2}\right) - \frac{v^2}{c^2} \quad (3)$$

This is Bethe- Bloch formula. This is valid for heavier charged particles only and is not applicable to electrons.

A plot of specific energy loss along the track of charged particles i.e., distance of penetration is known as Bragg curve. While passing through the matter, particles lose energy in small but definite amounts after undergoing a large number of collisions so that its energy is reduced by certain specific amount. There will always be statistical fluctuations in the amount of energy lost by particles having the same initial energy because the number of collisions by which each particle loses energy will be different. These fluctuation effects are known as Straggling. The property of straggling is common with all the heavy particles.

1.2.2. Interaction of Gamma with Matter

When radiation passes through matter it may interact with the material, transferring some or all of its energy to the atoms of that material.

Knowledge of the fundamentals of interaction of radiations with matter is important because it forms the basis of radiobiology and radiation protection. Interaction of radiation with matter depends upon many factors like mass, energy, atomic number, mass number and density of the medium.

The interaction of gamma rays with matter is different from that of charged particles. The difference is apparent in the much greater penetration power of gamma rays and in the absorption laws. Gamma rays are electromagnetic radiations having no definite range. Unlike charged particles, a well collimated beam of gamma rays shows the exponential absorption in matter, this is because photons are absorbed or scattered in a single event. If the radiation of intensity I is incident upon an absorbing layer of thickness dx, the amount of radiation absorbed dI is proportional both to dx and I so that

$$I = I_o e^{-\mu x} \tag{4}$$

where μ is the linear absorption coefficient.

There are number of processes which can cause gamma rays to be scattered or absorbed (Knoll G.F., 1999; Leo W.R., 1994). The main effects which are prominent at low energies are as follows:

1.2.2.1. Photoelectric Effect

The photoelectric process involves an interaction between an incident photon and bound electron. If the incident photon has low energy compared to the binding energy then classical scattering occurs. If however, the incident photon has energy just greater than the binding energy of the electron, a complete transfer of the photon energy to the electron occurs which results in an escape of electron.

A photon of energy hν ejects a K shell electron of binding energy (BE_k). The electron escapes from the atom with an energy E_k equal to that of the incident photon minus the energy required to release it from its orbit.

$$E_k = h\nu - BE_k \tag{5}$$

There is no scattered photon, all the energy has been transferred to the electron, known as photoelectron. Interactions of this kind can take place with electrons in the K, L, M, N shells.

1.2.2.2. Compton Effect

When radiations of shorter wavelength are used to irradiate materials of low atomic number, such as tissue, the electromagnetic radiation interacts with a free electron in the attenuating medium. A collision like process occurs in which the incident photon gives up some of its energy to the electron, known as Compton electron and the remainder of the energy appears as scattered photon of lower energy. In addition, the greater the angle through which the photon is scattered, the longer is the wavelength of the scattered radiation. This scattered radiation is called modified scatter and the process is referred to as an inelastic scattering or Compton scattering.

The proportion of photons scattered in the different directions varies with the energy of incident photon; low energy photons are scattered in all directions with almost equal probability, whilst high energy photons, greater than about 1 MeV, are scattered principally in the forward direction. The recoil electron behaves the same way as any electron. Since the recoil electron has been given kinetic energy, it can now undergo the various types of electronic particle interactions. It is indeed the recoil electron resulting from the Compton collision which causes the radiation damage to the tissue, not the initial photon interaction which has created only one ion directly. The Compton effect is independent of the atomic number.

1.2.2.3. Pair Production

Photons of energy greater than 1.022 MeV may interact with an electric field around the nucleus. This interaction results in the complete absorption of the photon and the simultaneous creation of two electrons, one positive and one negative. The production of these two electrons represents the creation of matter from energy. The equivalence of matter and energy was set out by Einstein in quantitative form as $E=mc^2$. In order to create two electrons, a minimum energy of 1.022 MeV is necessary. Since the electrons are opposite in charge, no creation of charge is involved in pair production. If the energy of initiating photon hν is greater than 1.022 MeV, then an excess energy will appear as kinetic energy E, shared between the electrons.

1.2.3. Interaction of Electron with Matter

Like heavy charged particles, electrons also suffer a collisional energy loss when passing through matter. However since electron has a small mass, so an additional energy loss mechanism comes into play, that is called radiative loss. Therefore, the total energy loss of electron is composed of two parts (Leo W.R., 1994).

$$\left(\frac{dE}{dx}\right)_{tot} = \left(\frac{dE}{dx}\right)_{rad} + \left(\frac{dE}{dx}\right)_{coll} \quad (6)$$

Actually the mode of interaction of electrons with matter depends upon their energies as follows:

a. When an energy of electron is small (eV to keV), it is scattered elastically by the cloud of electrons of atoms.
b. For slightly more energies, an electron penetrates the outer cloud of electrons of atom and undergoes inelastic scattering due to electrons.
c. With still higher energies, electron reaches close to the nucleus of atom and is scattered inelastically by the field of nucleus.
d. When the energy of electron is very high and is of the order of MeV, the electron interacts with the nucleus and results into an excitation of nucleus

So, it is observed that while passing through matter, electron loses their kinetic energy by two different ways:

(1) When electron energy is low, the loss due to ionization predominates and it may be treated in a similar manner to that of heavy

charged particles. In this case, inelastic collisions of incident electrons take place with the bound atomic electrons of target. These collisions result in an excitation and ionization of the atoms of matter through which they pass. The loss of energy per unit path length of an electron in low energy limit is given by the same relation as for charged particles i.e.,

$$\left(-\frac{dE}{dx}\right)_{ion} = \frac{e^4 n}{4\pi\varepsilon_0^2 m_e v^2}\log_e\left(\frac{2m_e v^2}{h\vartheta}\right) - 1.2329 \tag{7}$$

where Z= 1 for electron and there is an additional term (-1.2329). As

$$\left(-\frac{dE}{dx}\right)_{ion} \propto \frac{1}{E}$$

So, energy loss due to ionization is quite high at low energies and decreases as the energy increases.

When the energies of incident electrons are very large i.e., E>> 1 MeV, then electrons suffers radiative collisions with atomic nuclei of the material through which they pass. When a high velocity electron passes close to the nucleus of target atom, it experiences strong coulomb's force of attraction due to nucleus and is suddenly slows down and hence electron suffers a deflection in its path. During its retardation, an electron emits electromagnetic radiations which fall in X-ray region of electromagnetic spectrum. Due to different loses in velocity of incident electrons, a continous spectrum of X-ray is produced and this radiation with continous spectrum is called 'Bremstrahlung'. In this case the energy loss per unit path length of an electron is proportional to the square of atomic number Z of absorbing material i.e.,

$$\left(-\frac{dE}{dx}\right)_{rad} \propto Z^2$$

The loss of energy due to radiative collisions is more prominent in heavy metals.

1.2.4. Interaction of Neutron with Matter

Neutron, a neutral particle having no charge can be classified on the basis of their energy, although no specific boundaries are prescribed between the classes. Generally the neutrons with energies above 100 MeV or so are 'High energy neutrons' whereas those between a few ten's of MeV and few hundred keV are known as 'Fast neutrons'. Neutrons between energy range 100 keV and 0.1 eV where nuclear resonance reactions occur, are referred to as 'Epithermal neutrons'. At lower energies, comparable to thermal agitation energy at room temperature i.e ($E = kT = 1/40$ eV), neutrons are known as 'Thermal or slow neutrons'. Further the neutrons having energies of milli or micro eV are known as 'Cold or Ultra cold neutrons' (Knoll G.F., 1999).

Similar to photon, the neutron lacks an electric charge, so it is not subjected to the coulomb interactions with electrons and nuclei in matter, rather the interaction is through the strong force with nuclei. Actually neutrons require coming within $\approx 10^{-13}$ cm of nucleus before anything can happen. So, when the neutron interacts with matter, it may undergo a variety of nuclear processes depending on its energy that are explained below:

a. For slow neutrons, the significant interaction includes elastic scattering with absorber nuclei and a large set of neutron- induced nuclear reactions. Since the kinetic energy of neutrons is small, so very little energy can be transferred to the nucleus in elastic scattering.
b. If neutrons have sufficient energy to excite the nucleus, usually of the order of 1 MeV or more, an inelastic scattering occurs. In this reaction, the nucleus is left in an excited state which may later decay by gamma ray or some other form of radiative emission.
c. At low energies of neutron, sometimes radiative neutron capture reaction occurs. Generally, the cross section for neutron capture

goes approximately as ≈ 1/v, where v is the velocity of neutron. Radiative capture reactions can be useful in the indirect detection of neutrons using activation foils but are not applied in active neutron detectors, because secondary radiations takes the form of gamma rays which are also difficult to detect. Instead for neutron energies eV to keV range, other nuclear reactions occurs such as (n, p), (n, d), (n, α) etc in which neutron is captured and charged particles are emitted. These reactions are more attractive since secondary radiations are charged particles and can be easily detected.

d. At thermal energies of neutron, fission i.e., (n, f) occurs.
e. For very high energy of neutrons i.e., E > 100 MeV, high energy hadron shower production occurs.

So, the total probability for a neutron to interact with matter is given by the sum of individual cross sections

$$\sigma_{tot} = \sigma_{elastic} + \sigma_{inelastic} + \sigma_{capture} + \ldots \quad (8)$$

This quantity has the same significance for neutrons as the linear absorption coefficient for gamma rays.

1.2.5. Interaction of Proton with Matter

Protons are hydrogen nuclei and therefore are positively charged particles of mass approximately equal to that of neutron. Proton accelerated to high energy travel in straight lines, they are easily scattered and have precise range in tissue determined by their energy. Proton beam interacts with the matter in the same way as heavy charged particle interaction. A proton beam has a high dose “Bragg peak region” which is designed to cover the entire tumour in proton therapy for the treatment of cancer. Following the Bragg curve, an energetic protons, loses its energy progressively along its track, with most of its energy deposited near to where the particle comes to rest. Protons have linear energy transfer (LET) in the range from 5 to 100 keVμm^{-1} and therefore not very different from electrons. The dosimetry of proton beam is simpler in that there is less contamination of the proton beam with gamma radiation. Protons have three main advantages:

i. Their absorption characteristics in tissue are more favorable than photons.
ii. They have similar radiobiological characteristics to photons.
iii. The cost of delivering protons is less than that for heavier charged particles.

Moreover the most important physical advantage is that there is no dose delivered deep to the target volume for each beam path. So, the dosage to the normal structures outside the treatment volume is minimized. By contrast, in conventional photon irradiation, there is always an exit dose deep to the target volume, which may deliver substantial dose to normal tissues. Just as the high energy electron beam can be extracted from a linear accelerator by removing X-ray target, so the proton beam can be extracted from a cyclotron. The protons can be used for the treatment of cancer and the proton energy required to treat tumours at the depth, is very large.

1.3. Measurement of Ionizing Radiations

Chemical and biological changes in tissue exposed to ionizing radiation are caused by the deposition of energy from the radiation into the tissue. This deposition is described by two closely related quantities: kerma and absorbed dose. Both quantities are described in units of gray (Gy) or formerly, rads. A radiation dosimeter provides a measurable response to the energy absorbed in a medium from incident radiation. Different methods have been used to measure the radiation absorbed dose such as Ion chamber dosimetry, Semiconductor junction dosimetry, Solid state nuclear track dosimetry, Chemical dosimetry, Thermoluminescence dosimetry etc. In the present chapter, the focus is on thermoluminescence dosimetry.

1.3.1. Thermoluminescence

TL is a process in which light emerges from a heated material as a result of electrons (previously trapped within the material) being released. In other words, heat is used to free the electrons, which produce light. A very small fraction of the radiation energy can stay in an excited state for a long time. When such a material is heated, the stored energy is released as weak light or thermoluminescence. But earlier it was a misnomer and now correctly called as Thermally Stimulated Luminescence (TSL).

TL is the phenomenon of light emission on warming a previously irradiated substance with uniform heating rate. The TL phenomenon of minerals was known empirically as early as 1663. It was in this year that Sir Robert Boyle reported to the Royal Society about "experiments and considerations upon colors with observations on diamond that shines in the dark" and described how, upon warming a diamond in contact with his body in the dark, he saw a flash. This reflects the definition of TL. Probably, first time in 1895, the word thermoluminescene was published in literature (Wiedmann and Schmidt, 1895). This phenomenon was first applied by Hoffman in 1897 for the measurement of cathode rays, X- rays and ultraviolet (UV) radiation intensity by using calcium sulphate doped with manganese ($CaSO_4$: Mn) as a TLD material. The first application of this phenomenon for radiation dosimetry was given in 1953 (Daniels et al., 1953). TL is usually observed in ordered crystals, although the phenomenon also occurs in disordered materials such as glasses (Furtado et al., 1989; Del Nery et al., 1994; Venkateswara Rao et al., 2002; Pandey et al., 2004). TL phenomenon is widely used in the field of radiation dosimetry, but many problems related to TL sensitivity of the solid have been observed (Handee et al., 1968; Bartlett and Edwards, 1979; Suntharalingam and Cameron, 1969). TL is a very good technique used in research in basic as well as applied sciences. It is already applied in various fields and is now extended to a whole spectrum of disciplines such as archaeology, forensic sciences, geology, radiation dosimetry, radiation physics, solid state physics, space sciences and many more. There are number of commercially available thermoluminescent dosimeters, with their own advantages and disadvantages. Most popular are LiF: Mg, Ti (TLD-100), LiF: Mg, Cu, P (TLD-700 H), $CaSO_4$: Dy (TLD-900), CaF_2: Dy (TLD-200), Al_2O_3 (TLD-500) and many more will be available in near future.

1.3.2. Other Related Techniques

The techniques related to thermoluminescence can to applied to the same material, for example: thermally stimulated conductivity (TSC), thermally stimulated electron emission (TSEE), optically stimulated luminescence (OSL), photoluminescence (PL) etc and the combined information can yield a better picture of the studied sample. The mostly closely associated phenomenon to thermoluminescence is thermally stimulated conductivity (TSC). In this technique, electric contacts are formed on the sample, the sample is excited by ionizing radiation, after that an electric field is applied across the contacts of sample and sample is heated and during this heating, transient electric current is measured, usually in the form of peaks, similar in shape to TL peak. In phenomenon TL and TSC, the mobile charge carriers can be either electrons or holes, whereas in TSEE, only electrons take part. In this effect, sample is placed in a vacuum cryostat and is connected to an electrode. Another electrode in the tube is at certain distance from the sample, so potential difference is maintained between the two electrons. TL and TSC properties are related to bulk of the sample, whereas TSEE results are related to surface states.

Other effects are related to illumination of sample with light that are photoluminescence (PL) and optically stimulated luminescence (OSL). In this, the materials which have been exposed to radiation are illuminated with light and its OSL signal is recorded. The OSL intensity is then related to the absorbed dose to study the radiation dosimetry, whereas in case of photoluminescence, the light is emitted following the absorption of light by an internal transition. One more important related technique is Phototransfered TL (PTTL). It is a form of TL from shallow traps resulting from the transfer of charge from deeper traps. It is also a suggested tool for radiation dosimetry and in dating of ceramic samples.

1.4. Importance of Ionizing Radiations in Thermoluminescence Dosimetry

Thermoluminescence dosimetry is a technique used to make quantitative measurements of ionizing radiation exposure by measuring TL glow curves and TL responses to different radiations. There are number of TLD materials available commercially to be used for different applications. TL materials display varying TL responses to high energy photons and heavy charged particle beams (Kore et al., 2016). This variation is due to the differences in spatial dose distribution of both the radiations. Dosimetry of heavy ion beams is important due to its use in diagnostic and therapeutic applications (Schardt et al., 2010). The existence of Bragg peak region and greater relative biological effectiveness (RBE) due to high linear energy transfer (LET) of heavy ion beam makes them an important tool for cancer/tumour therapy. RBE is defined as the rate at which biological cells are killed by one type of ionizing radiation with respect to another having different radiation qualities to same amount of absorbed doses. The RBE value for heavy ion beam like carbon increases up to a particular value of LET while the RBE value of photon and proton beams does not change significantly as the LET increases (Schardt et al., 2010). So, there is maximum dose deposition at the Bragg peak region for heavy ion beam which is better than that of photon or proton beams (Schardt et al., 2010). Conventional radiation beams dissipates its energy throughout its path and thus provide normal tissue complication whereas HCPs deposits their maximum energy to a particular confined region of targeted volume with least scattering and negligible angular energy straggling effects.

Ionizing radiations are also used for modifying the optical and electronic properties of inorganic materials (Dutta et al., 2017). When ionizing radiations passes through materials, it loses energy via intense electronic excitations resulting in non-equilibrium conditions which help the system in achieving unique properties (Som et al., 2013). In the case of insulator or semiconducting phosphors, this inhomogeneous energy deposition may lead to the production of new color centers, and/or atomic vacancies (point defects) that modify the luminescence properties of materials (Som et al., 2013). TL is a well-known and very sensitive method for the characterization of such defects in solids.

1.5. Research Undertaken So Far

Today, varieties of materials have been identified for thermoluminescence dosimetry (TLD) with different efficiencies for different radiation dose ranges. Nanophosphors have potential role in many R&D areas like medical (Kron, 1999; Efstathopoulos et al., 2003), accidental (Veronese et al.,2010), retrospective (Kazakis et al., 2016; Mesterhzy et al., 2012), personal (Gai et al., 2015), thermal neutron (Bhadane et al., 2016) dosimetry, solid state lighting (Dalal et al., 2016; Naik etal., 2016) and 2D OSL mapping (Oliviera etal., 2016). There are number of standard commercial dosimeters available at international level and are well known by different names for their particular observed dose range. The most famous dosimeters developed are LiF: Mg, Cu, P (TLD-700H), Al_2O_3 (TLD- 500), $CaSO_4$:Dy (TLD-900), CaF_2:Dy (TLD-200) and many more may be available in the near future (Sahare etal., 2007; Fox et al., 1988; Noh et al., 2001). Most of the phosphors can be used as TLDs within a specific range of radiation doses and not for all doses from very low to very high range because it depends on various factors including linearity, precision, dose rate, fading, reproducibility and others. Researchers suggested different techniques to estimate high intense phosphor. In the last few decades, many new and useful TLD materials have been reported such as $Li_2B_4O_7$:Mn (Schulman et al., 1965), CaF_2:Dy (Binder et al., 1968), BeO (Tochilin et al., 1969) and Al_2O_3:Si, Ti (Mehta and Sengupta, 1976), CaF_2:Tm (Lucas and Kasper, 1977), LiF:Mg, Cu,P (Nakajima et al., 1978), $Li_2B_4O_7$:Cu (Takenga et al., 1980), MgB_4O_7:Dy or

Tm (Prokic, 1980) and Al_2O_3:C (Akselrod et al., 1990). Subsequently a large number of laboratories have prepared LiF:Mg, Cu, P; BeO; $CaSO_4$:Dy or Tm and some other TLD phosphors. Recently, many research groups are involved worldwide in exploring the best TLD for ionizing radiation dosimetry. Numan salah et al. studied the thermoluminescence properties of Al_2O_3:Tb nanoparticles (Salah et al., 2015) and LiF: Eu (Salah et al., 2015) irradiated by gamma rays and 85 MeV C^{6+} ion beam. Kulig et al. studied thermoluminescence and optically stimulated luminescence on $LiMgPO_4$ crystallized by micro pulling down technique (Kulig et al., 2016). The Prompt isothermal decay of thermoluminescence in MgB_4O_7:Dy, Na and LiB_4O_7:Cu, In dosimeters has been studied (Kitis et al., 2016). Further Zahidifar et al. of Iran studied the dosimetry features of $MgSO_4$:Dy and $MgSO_4$:Cu nano-rods (Zahidifar et al., 2016). A new thermoluminescence dosimeter, nano-barium–strontium sulfate is being reported (Aboelezz et al., 2015). Presently, researchers from different corners of the world are working on the thermoluminescence field to explore the sensitive materials best suited for radiation dosimetry. Synthesis and TL response of β-$Ca_2P_2O_7$:Tb^{3+} (Lopeza et al., 2017), LiF: Eu crystals (Vallejo et al., 2017), Mg_2SiO_4:Tb (Zhao et al., 2017), MgB_4O_7:Ce,Li for gamma and neutron dosimetry (Yukihara et al., 2017), NaSrBO3:Dy^{3+}/Sm^{3+} phosphor for 120 MeV Ag^{9+} and gamma dosimetry (Bedyal et al., 2017; Bedyal et al., 2017), $SrAl_2O_4$:Eu^{2+}, Dy^{3+} (Chithambo et al., 2017) has been reported by different groups internationally. Study of morphological and luminescent properties (TL and optical stimulated luminescence (OSL)) of ZnO nanocrystals has been reported (Soares et al. 2017). Further the dosimetric aspects of boron based nanocrystalline thermoluminescent phosphors: $Li_2B_4O_7$: Cu, MgB_4O_7: Dy, BCNO: Dy have been studied using gamma rays, proton beam of 3 MeV and 150 MeV and electron beam of 4 MeV and 9 MeV by our group (Chopra et al., 2014; Chopra et al., 2013; Salah et al., 2013; Singh et al., 2011; Singh et al., 2011; Chopra et al., 2012). Dhoble and group studied different materials e.g., $CaMg_3(SO_4)_4$ (CMS) (Kore et al., 2014), $CaWO_4$ (Dabrel and Dhoble, 2013), $KCaSO_4Cl$ (Kore et al., 2014), $Li_2BaP_2O_7$:Dy^{3+}(Wani et al., 2015), $LiCaAlF_6$:Ce (Yerpude et al., 2016), $LiCaBO_3$ (Oza et al., 2015),

Mg_2BO_3F:Dy (Kore et al., 2015), Na_3SO_4Cl:Ce (Shinde et al., 2014) and $Sr_5(PO_4)_3F$ (Gupta et al., 2016) phosphors for their suitability in the applications of gamma and carbon ion beam dosimetry. Further, the role of anion and cation vacancies in the thermoluminescence and photoluminescence processes of $BaSO_4$:Eu^{2+} has been studied (Rani and Lakshamanan, 2016). $CaSO_4$:Dy microphosphor has been studied for thermal neutron dosimetry (Bhadane et al., 2016). The TL sensitivity of $CaSO_4$:Dy,Mn phosphor was studied and was found to be about 2 and 1.8 times higher than that of standard phosphors: $CaSO_4$:Dy and LiF:Mg,Cu,P (TLD-700H) respectively (Bahl et al., 2016). Further synthesis and TL characteristics of MgB_4O_7:Mn,Tb phosphor has been studied (Sahare et al., 2015). Thermoluminescence, kinetic parameters of $SrSO_4$:Eu nanophosphors under X-ray and gamma excitations were studied (Jayasudha et al., 2016). Very recently thermoluminescence properties of orange-red emitting Eu^{3+} doped $Ca_2Al_2SiO_7$ phosphors have been studied (Tawari et al., 2017). A change in thermoluminescence behavior of cubic Gd_2O_3:Yb^{3+} phosphors with successive increase in Yb^{3+} ion concentrations have been noticed (Tamrakar et al., 2017). Rawat et al. studied sub micro gray dose levels using OSL phosphor $LiMgPO_4$:Tb,B for its application in personnel and environmental monitoring and nuclear forensics (Rawat et al., 2017). So in the light of above review it is seen that scientific interest is prevailing worldwide for the preparation of various materials for dosimetric applications and the importance of different ionizing radiations: gamma, electron, proton, neutrons, heavy ions for thermoluminescence dosimetry that is discussed in the following sections:

2. Thermoluminescence Dosimetry of Gamma Irradiated Materials

Thermoluminscence is a powerful technique to estimate the dose of high energy ionizing radiation, absorbed during exposure. A number of commercially available thermoluminescent dosimeters are available for this purpose (Fox et al., 1988; Noh et al., 2001). Since the basic requirement for TLD is that it should have high sensitivity at low & high doses, linear TL response over wide range of doses, low fading, good thermal stability, energy independence & reproducibility. So, it is challenging for researchers to develop high quality phosphors, having all these properties. Recently, nanocrystalline phosphors have found their applications as TLD materials, since they are found to have improved TL characteristics. Basically, physical properties of individual nanoparticles can be very different from those of bulk materials. Nanomaterials are found to be more stable, they do not get deteriorated at high doses of high energy radiations, their glow curve structure do not change with dose and do not saturate even at very high doses when microcrystalline phosphors saturate (Salah et al., 2006; Salah et al., 2006; Sahare et al., 2007; Lochab et al., 2007; Lochab et al., 2007; Salah et al., 2009). So, now a days research is mainly focused on developing nanomaterials as dosimeters. In the past years, only Co-60 therapy was used for cancer treatment in the hospitals, hence only gamma ray dosimeters were explored, but later the ion beam therapy came into scene because of its advantage over Co-60 therapy and ion beam dosimeters have been explored. Instead of medical, TLD's are being used in different fields such as environmental monitoring, personal dosimetry, biology, neutron dosimetry, dating, space research etc (Bhadane et al., 2016).

2.1. Gamma Ray Dosimeters

TL dosimeters are sometimes classified on the basis of their effective atomic number Z_{eff}. If Z_{eff} is less than 16, it is considered as low Z_{eff} TLD, which is commonly used for low dose applications below 1 kGy whereas materials with Z_{eff} greater than 16 are considered as high Z_{eff} TL dosimeters, used for high dose applications with dose greater than 1 kGy. Currently, commercially available thermoluminescent dosimeters are LiF: Mg, Ti (TLD-100), LiF: Mg, Cu, P (TLD-700 H), $CaSO_4$: Dy (TLD-900), CaF_2: Dy (TLD-200), Al_2O_3 (TLD-500) and many more will be available in near future (Fox et al., 1988; Noh et al., 2001). Efforts are still being made to improve TL properties of these phosphors by synthesizing those using different methods, doping with different impurities & giving them different heat treatments (Salah et al., 2006; Salah et al., 2006; Sahare et al., 2007). Few of the gamma ray dosimeters are explained below:

2.1.1. Commercial Dosimeters: LiF & CaSO4: Dy

LiF: Mg, Cu, P is highly sensitive TLD material and is one of the most studied sample and still work is in progress to improve its properties (Shinde et al., 2001). Presently this phosphor is available in many forms, such as microcrystalline powder, chips, sintered disks & microrods (Shinde et al., 2001). Recently, the dosimetric characterstics of LiF: Mg, Cu, P nanocrystalline powder has been studied (Salah et al., 2007). Nanocrystalline powder of LiF: Mg, Cu, P was prepared by co-precipitation method (Salah et al., 2006) and was studied for its TL properties. TL glow curve for nanocrystalline LiF: Mg, Cu, P powder exposed to 100 Gy of γ-rays from a ^{137}Cs source was studied along with the TL glow curves of standard phosphors LiF: Mg, Ti (TLD-100), LiF: Mg, Cu, P (TLD-700 H) for comparison purpose.

It was observed that the nanomaterial powder has single TL glow peak at 410 K with broad peak at around 576 K whereas TLD-700H phosphor shows a dosimetry peak at around 490 K. Moreover, the sensitivity of nanomaterial powder was found to be 12 times less than that of TLD-700H whereas it is 3 times more than the sensitivity of TLD-100. Further the response curve of TLD-700H saturates at around 1 kGy while that of nanocrystalline powder still increase upto 10 kGy & then saturates at around 20 kGy. So, the nanocrystalline phosphor could be used for estimation of very high doses of gamma rays.

Another commercial dosimeter $CaSO_4$: Dy for very high doses was explored for its TL properties. The TL glow curve for microcrystalline phosphor exposed to 10 Gy of γ-rays from ^{60}Co source was observed. It was found to consists of prominent dosimetry peak at around 494 K along with two small peaks at 406 & 594 K and small shoulder at around 673 K whereas the glow curve for nanocrystalline was very different than microcrystalline phosphor, consisting of peaks at around 411, 498, 569 & 652 K (Salah et al., 2006). TL sensitivity of nanocrystalline material was found to be less than that of microcrystalline phosphor at low & medium doses where as sensitivity was found to be much better for nanophosphor at high doses.

2.1.2. $K_2CaMg(SO_4)_3$: Eu Phosphor

Mixed sulphate $K_2CaMg(SO_4)_3$: Eu (KCM) was synthesized using solid state diffusion method (Kore et al., 2016) and was found to have same sensitivity as commercial $CaSO_4$:Dy. The glow curves for 15 Gy γ-rays exposed KCM for different concentrations of Dy are shown in Figure 1.

The glow curves for KCM shows two peaks, One lower temperature peak at 392 K and another higher temperature peak that further consists of two peaks at 498 K and 538 K. Further commercial CaSO4:Dy was also irradiated at the same dose for comparison purpose and it was observed that CaSO4: Dy shows same intensity as KCM phosphor with 0.5 mol % of Dy concentration. Further TL response of KCM phosphor for wide γ dose range from 15 Gy to 50 kGy is shown in Figure 2.

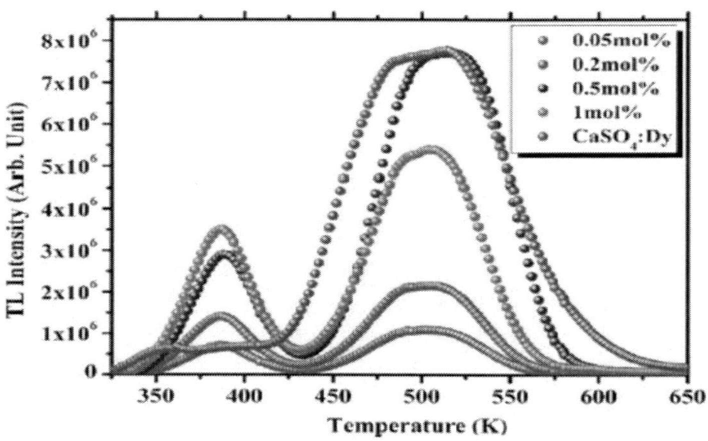

Figure 1. TL glow curves of γ-ray irradiated KCM phosphor with different Dy concentrations for 15 Gy dose (Reproduced with permission from ref. (Kore et al., 2016), Copyright 2016, IOP publishing Ltd).

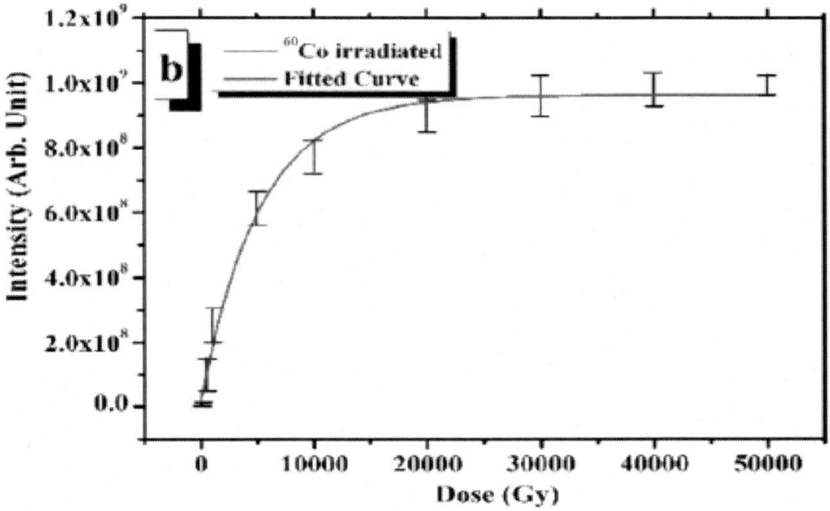

Figure 2. TL dose response of KCM phosphor irradiated with γ-rays from ^{60}Co source (Reproduced with permission from ref. (Kore et al., 2016), Copyright 2016, IOP publishing Ltd).

It is observed that KCM phosphor shows linear behaviour for 10 kGy and above this dose, the material shows saturation. So, the dose upto 10 kGy will be beneficial from application point of view.

2.1.3. $Li_2B_4O_7$: Cu and MgB_4O_7:Dy Phosphors

Borates are of considerable interest in TL dosimetry because borates are chemically stable and do not face any problems while doping. Materials such as $Li_2B_4O_7$: Cu and MgB_4O_7:Dy have a close tissue equivalence and thus worth considering for their TL properties. Microcrystalline $Li_2B_4O_7$: Cu and MgB_4O_7:Dy have already been synthesized using different methods and their TL response was studied in details for their applications in dosimetry. Recently $Li_2B_4O_7$: Cu and MgB_4O_7:Dy have been synthesized using combustion method (Singh et al., 2011; Chopra et al., 2013; Lochab et al., 2007) and TL properties have been studied. Figure 3 & Figure 4 (a), (b) shows TL response of nanocrystalline $Li_2B_4O_7$: Cu and nanocrystalline MgB_4O_7:Dy for different gamma doses.

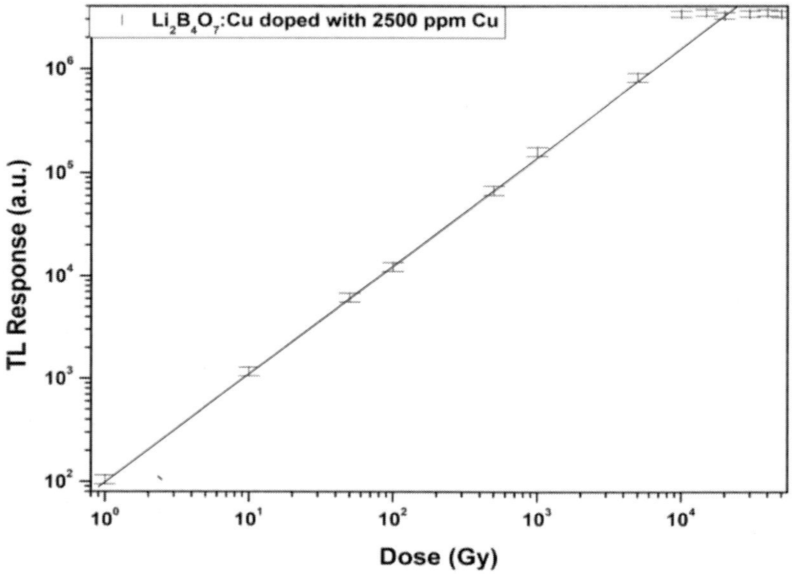

Figure 3. TL response of nanocrystalline $Li_2B_4O_7$:Cu doped with 2500 ppm Cu for wide range of γ- doses (Reproduced with permission from ref. (Chopra et al., 2013), Copyright 2016, Elsevier).

It is observed from Figure 3 that $Li_2B_4O_7$: Cu doped with 2500 ppm Cu exhibits a linear response from $1x10^0 - 5x10^3$ Gy of gamma dose. The TL response of the synthesized MgB_4O_7:Dy at low doses shown in Figure 4 (a) shows a remarkable linear pattern at lower doses in the range $4x10^{-4}$ to $4.5x10^{-2}$ Gy. When dose is increased to $1x10^{-1}$ Gy, the material shows sublinear behavior. This shows that the synthesized MgB_4O_7:Dy nanophosphor can be used as dosimeter for low doses in the range $4x10^{-4}$ to $4.5x10^{-2}$ Gy. Further the TL response of MgB_4O_7:Dy in the range $1x10^{-1}$ to $5x10^4$ Gy (Figure 4 (b)) reveals that the synthesized nanophosphor MgB_4O_7:Dy exhibits a sublinear behaviour below $1x10^0$ Gy dose and becomes linear in the range $1x10^0$ Gy to $5x10^3$ Gy. With further increase of dose, the material shows saturation. These results were found to be in consistent with earlier reported results (Lochab et al., 2007). So, both the nanophosphors may be used for high dose measurements of gamma radiations.

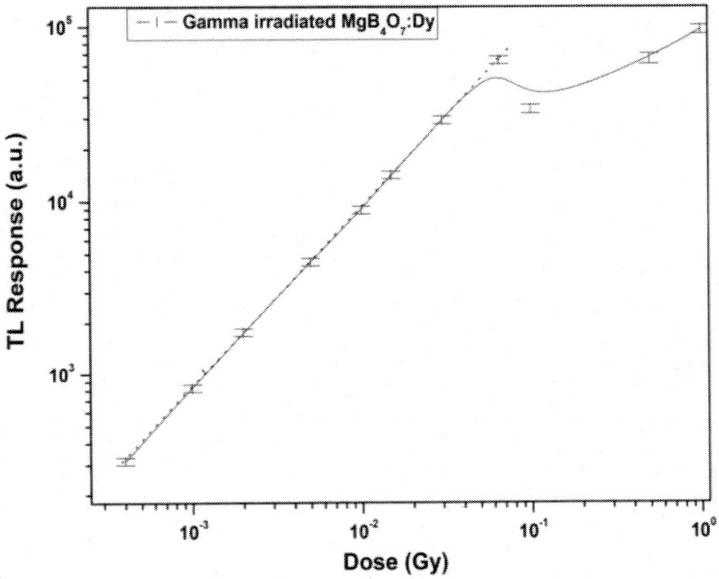

(a)

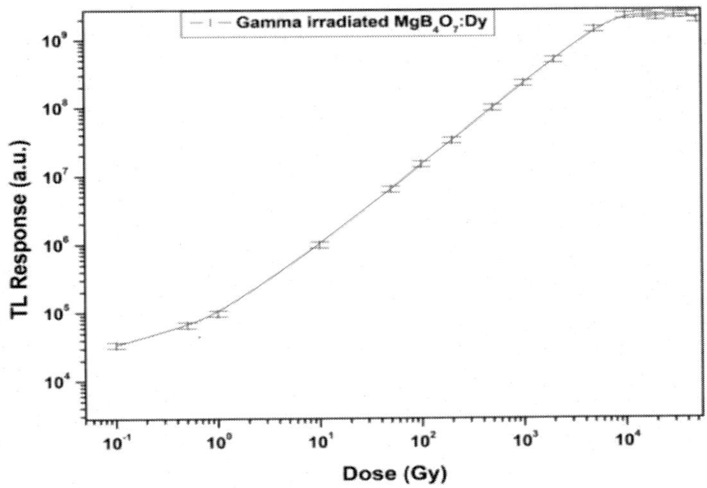

(b)

Figure 4. TL response of MgB_4O_7:Dy nanophosphor for (a) low γ-doses (b) high γ-doses.

3. Thermoluminescence Dosimetry of Electron Irradiated Materials

High energy electron beams have been largely applied in radiotherapy, where their application demands a great accuracy and precision in the delivered dose because a variance of 5% in the absorbed dose by the tumour can cause reoccurrence risk. So, dose measurement by dosimeters needs to be highly accurate (Nunes and Campos, 2008). Now the absorbed dose distribution obtained in electron beam normally depends on the design parameters as compare to photon beam. Electrons are more influenced by scattering and energy loss interactions with the materials than photon does (De Almeida & Almond 1974 a; Brahme etal., 1975; Brahme & Svensson, 1979). Moreover gamma and electron radiation exhibits different dosimetric properties due to different dose distributions. However, electron sources offers short treatment time and very low stray radiation that seems to be ideal, especially for treating coronary tumours. Further electron sources used for dosimetry are easy to handle as very simple & small amount of shielding can stop electrons.

3.1. Electron Dosimeters

Electron beam irradiation has become a potential method in basic & applied sciences to modify the properties of materials (Burkert et al., 2009). Electron beam has high penetration depth and hence a sufficient energy to cause damage in the material due to secondary electrons produced (Briand et al., 1990). These defects created due to collisions of atoms in the material makes the material important for radiation dosimetric applications. Different materials like LiF; LiF:Mg, Ti; $CaSO_4$: Dy; Li_2B_4O7: Cu; MgB_4O_7: Dy; ZnO; Y_2O_3: Sm; Y_2O_3: Tb; Al_2O_3 etc has been studied for electron dosimetric applications, few of them are explained as follows:

3.1.1. Commercial Dosimeters: LiF & $CaSO_4$: Dy

TLD plays an important role in the radiation therapy for electron dosimetry, majority of measure being done by lithium fluoride, LiF (Gooden & Brickner, 1972; Ruden, 1976) as a base material. LiF, in the form of LiF: Mg, Ti-TLD 100 has been used in TLD industry since many years and is the most widely used TLD material (Mandowska et al., 2002). Both pure and Mg, Ti doped LiF shows excellent TL response linearity. However another phosphor $CaSO_4$: Dy has been used commercially due to its high sensitivity.

The high purity LiF pallets were produced using crystals grown by Czochralski method (Ribeiro et al., 2008). The pallets were annealed in air at 400°C for 1 h and then after cooling they were irradiated to different doses of electron beam at energies 8 MeV, 10 MeV & 14 MeV. The TL response of a pure LiF detector exposed to electron beam of different energies was noticed. It was observed that TL responses are linear in the dose range from 0.1 to 2 Gy and the TL sensitivity is inversely proportional to electron energies.

Further this material was explored by doping and the TL properties of LiF:Mg, Ti were studied along with another phosphor $CaSO_4$: Dy. Dose response curves of LiF: Mg, Ti to electrons of energy 4, 6, 9, 12 & 16 MeV at the depth of maximum dose in SW solid water phantom over 5 cm (Nunes and campos, 2008) were studied. It was found that the responses of LiF:Mg, Ti currently used in applications of radiotherapy with dosimetric purposes and $CaSO_4$: Dy TL dosimeters to electrons are linear for doses ranging from 0.05 to 1.25 Gy. Moreover, $CaSO_4$: Dy sensitivities are found to be 17.8- 21.5 times greater than LiF: Mg, Ti sensitivity for electron energies.

3.1.2. $Li_2B_4O_7$: Cu and MgB_4O_7:Dy Phosphors

Nanocrystals of $Li_2B_4O_7$: Cu were synthesized using combustion method for 1000 ppm and 2500 ppm of Cu. The synthesized samples were exposed to electron beam of 4 MeV & 9 MeV energies for different doses (Chopra et al., 2014) and their TL characteristics were studied. The results are shown in Table 1.

Table 1. TL properties of synthesized $Li_2B_4O_7$: Cu (doped with 1000 ppm and 2500 ppm Cu), irradiated with electron beams of different energies

Electron beam Energy	Observations	Results of LiB_4O_7: Cu (1000 ppm)	Results of LiB_4O_7: Cu (2500 ppm)
4 MeV	Exposure range	$1x10^{-2} – 1x10^{1}$ Gy	$1x10^{-2} – 1x10^{1}$ Gy
	Peak position for highest dose	Prominent peak – 482 K Small Peak- 609 K	Prominent peak – 465 K Small hump- 514 K Small Peak- 609 K
	TL Response Linearity	$1x10^{-1} – 1x10^{1}$ Gy	$1x10^{-1} – 1x10^{1}$ Gy
	Fading	7.2% in month	% in month
9 MeV	Exposure range	$1x10^{-2}$ - $1x10^{1}$ Gy	$1x10^{-2}$ - $1x10^{1}$ Gy
	Peak position for highest dose	Prominent peak - 475 K Small hump- 540 K Small Peak- 615 K	Prominent peak - 461 K Small hump- 511 K Small Peak- 592 K
	TL Response Linearity	$1x10^{-1} – 1x10^{1}$ Gy	$1x10^{-1} – 1x10^{1}$ Gy
	Fading	7% in month	7.3% in month

It was observed that $Li_2B_4O_7$: Cu doped with Cu (1000 ppm & 2500 ppm) shows a linear TL response in the range $1x10^{-1} – 1x10^{1}$ Gy for both 4 MeV & 9 MeV electron beam. So, $Li_2B_4O_7$: Cu nanophosphor was found to be promising candidate to be used in electron dosimetry.

Further TL properties of synthesized MgB_4O_7: Dy (Salah et al., 2013; Lochab et al., 2007), irradiated with electron beam of energy 4 MeV & 9 MeV for doses in the range $1x10^{-1} – 1x10^{1}$ Gy have been studied. TL response curves for both 4 MeV & 9 MeV energy irradiated MgB_4O_7: Dy are shown in Figure 5 and Figure 6.

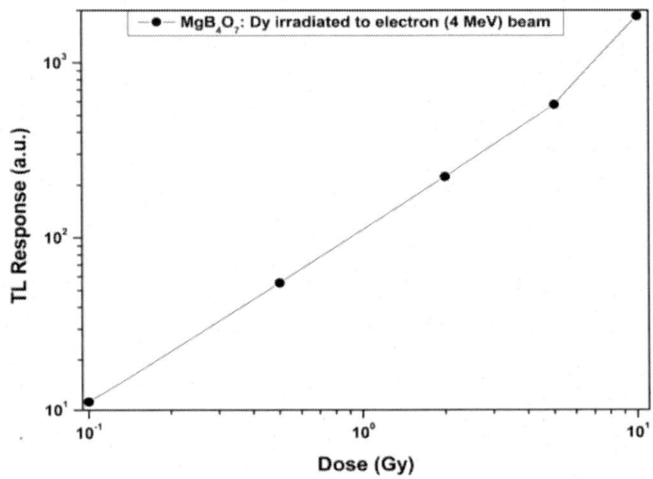

Figure 5. TL response of MgB_4O_7:Dy irradiated to electron (4 MeV) beam.

It has been found that MgB_4O_7: Dy nanophosphor irradiated to 4 MeV electron beam exhibits a linear response in the range $1x10^{-1} – 5x10^0$ Gy and for 9 MeV electron beam, linearity is in the range from $1x10^{-2} – 2x10^0$ Gy with the fading of almost 7% in both the cases. So, MgB_4O_7: Dy nanophosphor was found to be competent candidate to be used as electron dosimeter.

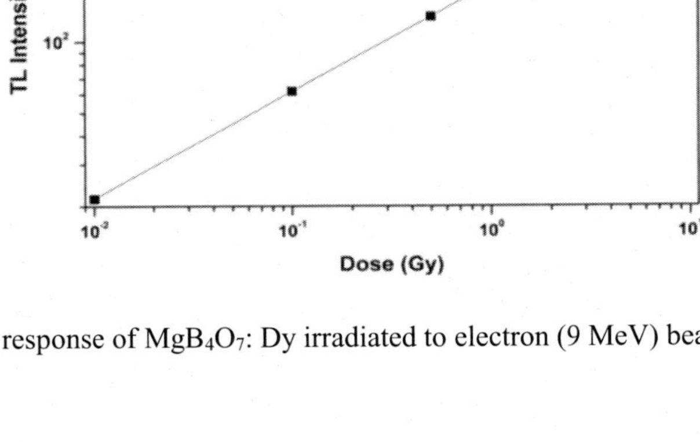

Figure 6. TL response of MgB_4O_7: Dy irradiated to electron (9 MeV) beam.

3.1.3. KMgF3:Sm Phosphor

Fluoro Perovskite doped with samarium ($KMgF_3$: Sm) has been synthesized by using Kiropoulos technique (Patterson, 1962) and TL properties were studied after irradiating the samples with different doses of electron beam (Marcazzo et al., 2010). It was observed that $KMgF_3$: Sm doped with 0.020 mol % samarium shows maximum TL intensity. TL glow curve was found to be made up of five peaks out of which third peak was best suited to be applied in personal & environmental dosimetry since it shows linearity in the dose range 0.02-200 Gy and no fading at room temperature.

4. Thermoluminescence Dosimetry of Proton Irradiated Materials

Proton radiation therapy is a form of external beam radiation treatment, radiation oncologists treat in two major ways. One way is by implanting radioactive materials in the body. The other is by using machines to generate beams that penetrate the body from outside. Proton radiation is delivered by the second method. Protons, a positively charged subatomic particles, deposits energy differently than X-ray beam & gamma ray. Compare to gamma or X-ray, the proton beam has a low entrance dose and high dose 'Bragg peak' region (Podgorsak, 2006), which is designed to cover the entire tumor and there is no exit dose beyond the tumour involved in proton radiation therapy (Newhauser & Zhang, 2015). So, the proton beam allows the maximum dose to be confined to the treatment volume, while the dose to the surrounding normal tissues is minimized (Devicienti et al., 2010; Lawrence and Feng, 2013). R. R. Wilson has first introduced protons for radiotherapy treatments in 1946 (Wilson, 1946). Moreover the cosmic rays near low earth orbit consists mainly of proton beams whose dosimetry becomes a challenging task. There are number of commercial TL detectors available but they have their own disadvantages, so efforts are still being made continuously to find the better TLD material for proton therapy.

4.1. Proton Dosimeters

Radiotherapy with proton beam is one of the most modern technique of cancer treatment. The basic dosimetric tool in proton therapy is ionization chamber. However thermoluminescence detectors based on Lithium fluoride: LiF: Mg, Ti; LiF: Mg, Cu, P are widely exploited for proton dose measurements now a days. Further research is going on for getting more and more sensitivity for proton beam using different materials like $CaSO_4$, $BaSO_4$, $K_2Ca_2SO_4$, $K_2Ca_2(SO_4)_3$, etc. Few of them are discussed below:

4.1.1. Commercial Dosimeters: LiF & $CaSO_4$: Dy

Commercial dosimeter LiF: Mg, Ti and LiF: Mg, Cu, P in the form of sintered pallets were procured and were annealed at 400°C for 1 h followed by 100°C at 2 h and 240°C at 10 min respectively. Then the samples were irradiated to different energies of proton beam and their TL properties were studied (Sadel et al., 2015). The normalised TL glow curves for both materials (LiF: Mg, Ti; LiF: Mg, Cu, P) for proton & gamma rays have been noticed and it is observed that for 58 MeV & 100 MeV protons, glow curves do not change, but for energy 20 MeV, change in glow curve is observed especially for high temperature region. The measurements were performed at different proton facilities and results were found to be consistent for all.

Further the thermoluminescent properties of $CaSO_4$: Dy (TLD-900) has been studied (Mathur et al., 1999). $CaSO_4$: Dy samples were irradiated with 3 MeV proton beam and TL response after a proton dose of 50 and 500 kGy was recorded and it was observed that two high temperature peaks in 335°C and 475°C were shown in the glow curve for each dose. It is further noticed that peak at 335°C saturates at about 10 kGy but the peak at 475°C saturates beyond 50 kGy. So, the use of high temperature thermoluminescence peaks makes this material quite suitable for use in reactor or hot waste dosimetry.

4.1.2. $K_2Ca_2(SO_4)_3$: Eu Phosphor

The mixed sulphate $K_2Ca_2(SO_4)_3$: Eu was prepared by solid state diffusion method (Sahare and Mohril, 1990) and the thermoluminescent response of synthesized material to 150 MeV proton beam for dose range from 0.1 Gy to 300 Gy has been reported (Bahl et al., 2012). It has been found that the phosphor shows a single peak at around 420 K. The TL response is found to be linear in the range upto 200 Gy. The linear TL response of $K_2Ca_2(SO_4)_3$: Eu with low fading & good reusability makes it a potential candidate to be used as a dosimeter for detecting the doses of proton beams for specific applications.

4.1.3. $Li_2B_4O_7$: Cu and MgB_4O_7:Dy Phosphors

Borates are of considerable interest in TL dosimetry. Materials such as $Li_2B_4O_7$: Cu and MgB_4O_7:Dy have a close tissue equivalence and are thus worth considering for their TL properties. So, $Li_2B_4O_7$: Cu doped with 2500 ppm Cu and MgB_4O_7:Dy doped with 1000 ppm Dy were synthesized by using combustion method. TL properties for both were studied using 150 MeV proton beam at different doses by our group and are summarized in Table 2.

Table 2. TL properties of synthesized $Li_2B_4O_7$: Cu and MgB_4O_7: Dy irradiated with 150 MeV proton beam

Beam	Observations	Results of MgB_4O_7:Dy	Results of LiB_4O_7: Cu
150 MeV Proton beam	Dose range	$1x10^0 – 1x10^2$ Gy	$1x10^0 – 1x10^2$ Gy
	Peak position for highest dose	Prominent peak – 412 K Small Peak- 509K	Prominent peak – 443 K Small Peak- 492 K, 550 K
	TL Response	Lineari ty- $5x10^0 – 1x10^2$ Gy	Linearity- $1x10^0 – 3x10^1$ Gy
	Fading	8.2% in month	% in month

It has been found that $Li_2B_4O_7$: Cu show a linear TL response for dose 1 Gy to 30 Gy of proton beam of energy 150 MeV whereas MgB_4O_7: Dy shows linearity in the dose range 5 Gy to 100 Gy. So, both the materials are found to be competent enough to be used as proton dosimeters.

5. Thermoluminescence Dosimetry of Heavy Ion Irradiated Materials

The demand of dosimetry of charged particles is increasing now days due to its utility in cancer diagnosis and therapy (Barth et al., 2001; Barth et al., 2003; Haberer, 2002). Ion beam therapy is superior tool in clinical practice for curing cancer in comparison to conventional radiation beam (Eickhoff et al., 1999; Amaldi and Kra, 2005). Heavy charged particles such as C^{5+} ions are heavier than the constituent particles in conventional radiation that ensure the penetration of radiation into the final treatment location deep inside the body with minimum scattering and stiffer particle trajectories, low straggling effect and sharper field edges. Moreover, ion beams have well defined range and small angular scattering, favourable dose depth distribution compared to photon or electron beam therapy. Carbon ion beam therapy exhibits more significant biological effect than other types of therapy whereas a heavier and lighter ion possesses their own drawbacks. The limitations of using heavier ion such as Ne & Ar is that they cause irreparable damage at the entrance, so damage healthy tissues around tumour, whereas in light ions such as protons, the scattering effect is large (Kore et al., 2014). Currently the importance of nanomaterials in the field of thermoluminescence has been increased as they exhibit enhanced properties, so research community is in continuous search of best dosimeter for ion beams.

5.1. Heavy Ion Beam Dosimeters

TL dosimeter materials for heavy ion beam presently in use are inorganic crystalline materials due to their ability to emit visible light radiation when suitably excited. There are several studies on different TLD materials aimed at studying their TL response/efficiencies to different ion beams (Yang et al., 2002; Bilski et al., 1997; Bilski et al.1998; Bilski et al.1994; Sahare et al., 2005; Salah et al., 2006; Salah & Sahare 2006; Lochab et al., 2007). However most of these TLD materials are not capable of precisely detecting HCP. The saturation effect occurs at very low and very high fluence which is undesirable for dosimetry. Recently highly sensitive nanophosphors have been synthesized for their applications in heavy ion beam dosimetry such as $CaSO_4$: Dy; LiF: Mg, Cu, P; $K_2Ca_2SO_4$: Eu and $Ba_{0.97}Ca_{0.03}SO_4$: Eu (Salah 2008; Salah et al., 2007; Salah et al., 2008; Lochab et al., 2008). The important results obtained in some of these nanomaterials and the possibilities of using them as HCP dosimeters have been discussed below:

5.1.1. Commercial LiF & $CaSO_4$:Dy Dosimeters

Nanocrystalline LiF: Mg, Cu, P of rod shape (about 30-40 nm) has been prepared by chemical co-precipitation method (Salah et al., 2007). The synthesized nanorods were irradiated by 48 MeV ^{7}Li ion beam at different doses in range 0.679-679 kGy (Salah et al., 2008; Salah, 2011). The TL glow curve shows prominent peak at around 588 K and smaller one around 410 K. TL response cure of LiF: Mg, Cu, P nanorods exposed to different fluencies of 48 MeV ^{7}Li beam was noticed. The linearity of response curve for the fluence range $1x10^{11}$- $1x10^{13}$ makes it a suitable candidate to be used as dosimeter in the respective range of heavy ion beam.

Further nanocrystalline form of $CaSO_4$: Dy was prepared (Salah et al., 2006) and the samples were irradiated at room temperature to C^{6+} ion beam at energy 75 MeV for ion fluencies in range $1x10^9$-$1x10^{13}$ ions/cm^2 (Salah et al., 2008). It has been seen from TL response curves of $CaSO_4$:Dy nano and microcrystalline samples to 75 MeV C^{6+} ion beam that the microcrystalline $CaSO_4$:Dy is linear in the range $1x10^9$-$1x10^{11}$ ions/cm^2 whereas nanocrystalline phosphor shows linearity in the fluence range $1x10^9$-$1x10^{13}$ ions/cm^2 (Salah et al., 2011). This shows that nanocrystalline form of $CaSO_4$:Dy is useful for detecting the high doses of carbon ions used in radiotherapy.

5.1.2. $K_2Ca_2(SO_4)_3$: Eu Phosphor

The microcrystalline $K_2Ca_2(SO_4)_3$: Eu was found to be two times more sensitive than TLD-700H for ɤ-rays (Sahare and Mohril, 1990). TL sensitivity of $K_2Ca_2(SO_4)_3$: Eu microphosphor was found to be more than that of TLD-100 (LiF: Mg, Ti) but quite less compared to TLD-700H (Pandey et al., 2002 a). Later $K_2Ca_2(SO_4)_3$: Eu nanoparticles were prepared and were exposed to three different ions (Li, O, C) having different energies. The fluence range was $1x10^9$-$1x10^{13}$ ions/cm^2. The TL glow curves of nanomaterials irradiated by Li^{3+}, C^{6+} and O^{7+} ion beams were found to be similar with single peak in each at around 427 K. It is observed that the TL intensity of irradiated nanomaterial decrease while moving from low to high atomic number for ions (Li→O). Moreover TL response curve is linear for whole range for Li^{3+} ions whereas for C^{6+} & O^{7+} the linearity is upto $1x10^{12}$ ions/cm^2. So, this makes the nanomaterial $K_2Ca_2(SO_4)_3$: Eu a suitable candidate for detecting the doses of swift heavy ions.

5.1.3. Mg_2BO_3F:Dy Phosphors

Mg_2BO_3: Dy phosphor was synthesized using wet chemical method to produce single phase (Kore et al., 2015). The samples were irradiated by C^{5+} ion beam at 50 MeV and 75 MeV energies for different ion fluencies. Figure 7 shows the TL glow curves for 50 MeV and 75 MeV energies C^{5+} ion beam, γ-ray irradiated Mg_2BO_3F: Dy and $CaSO_4$: Dy.

TL glow curves for C^{5+} irradiated samples shows two prominent peaks, and the glow curve structure of C^{5+} ion and γ irradiated samples seems to be similar. Further the intensity was found to increase from 50 MeV to 75 MeV.

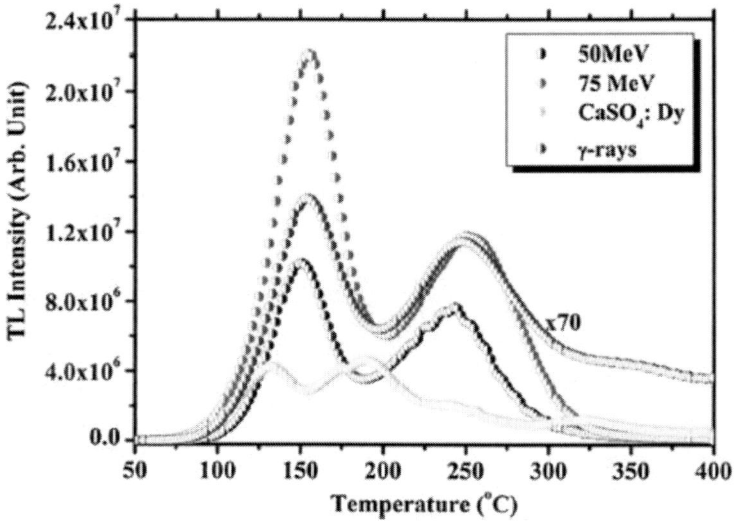

Figure 7. TL glow curves of C^{5+} irradiated Mg_2BO_3F:Dy sample at 50, 75 MeV with $CaSO_4$:Dy TLD phosphor for comparison (Reproduced with permission from ref. (Kore et al., 2016), Copyright 2015, Elsevier).

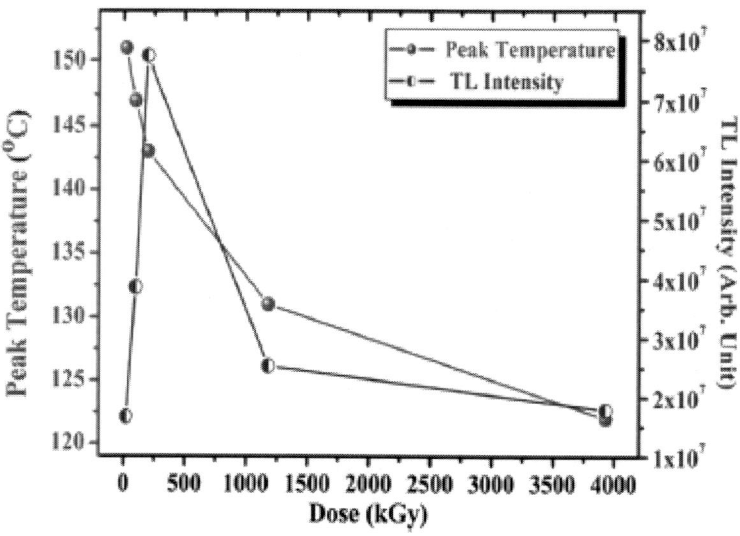

Figure 8. Variation of TL intensity and glow peak temperature of Mg_2BO_3F:Dy phosphor with 75 MeV C^{5+} ion beam in the dose range 198 kGy-3930 kGy (Reproduced with permission from ref. (Kore et al., 2016), Copyright 2015, Elsevier).

Moreover the variation of TL intensity and glow peak temperature of Mg_2BO_3F:Dy phosphor with 75 MeV C^{5+} ion beam in the dose range 198 kGy-3930 kGy is shown in Figure 8.

It is observed that the glow peak intensity increases supralinearly upto 197 kGy and thereafter it goes on decreasing. The supralinear region is attributed to the production of large number of traps inside phosphor resulting in higher TL intensity (Chen and Fogel, 1993; Larsson and Katz, 1976; Sunta et al., 1994). Strong TL properties make this phosphor important for dosimetry of carbon beams.

6. THERMOLUMINESCENCE DOSIMETRY OF NEUTRON IRRADIATED MATERIALS

Growth in the nuclear power production, development in the use of neutrons for radiotherapy has increased the number of radiation workers those must be monitored for neutron exposure. To develop an accurate neutron dosimeter is challenging since fast neutrons are considered to have a higher biological effectiveness (RBE) than an equal dose of γ-rays (Lakshmanan, 1982). Three types of personal neutron dosimeters have been used; Personnel neutron film type NTA (Nuclear track emulsion type A), Fission track detectors and TL albido dosimeters. NTA technique relies on detection of tracks in a thin photographic emulsion. The tracks are counted manually, but NTA faces the problem of fading. Further use of threshold detectors such as ^{237}Np, ^{232}Th or ^{238}U with polycarbonate fission fragment detectors provide us with fast neutron dosimeter that does not fades as NTA films does. But the use of ^{237}Np is prohibited for personal dosimetry. To overcome this problem, the new fairly sensitive Albedo dosimeters came into existence (Hankins, 1973). The simple albedo dosimeter consists of a pair of ^6LiF & ^7LiF TLD's on each side of cadmium and boron shield. ^6LiF has a very high TL sensitivity to thermal neutrons & both ^6LiF & ^7LiF have almost equal sensitivity to gamma rays. The results of albedo neutron dosimeter can be misinterpreted if gamma dose is high compared to neutron dose, because the readings of ^{7}LiF are subtracted from reading of ^{6}LiF TLD. Finally the new detector was developed, that is widely used now-a days: CR-39 plastic detector (Cassoll and Benton, 1978). This does not require fissionable radiators. It registers recoil protons from the plastics, which leaves damage tracks that can be revealed by various etching methods.

6.1. Neutron Dosimetry Materials

Personal neutron dosimetry has been pursued more rapidly for the safety of workers in neutron environment. Earlier till the year 1990, only Kodak NTA emulsion packs were being used for this purpose, but due to drawbacks of high neutron detection threshold energy of 500 keV, gamma sensitivity, eye straining job of evaluation of tracks, it was necessary to have another neutron personal monitoring system. CR-39 solid state track detector with 100 keV energy threshold and nearly flat response with energy was found to be suitable detector (Lakshmanan, 1982; Dhairyawan et al., 2003; Castillo et al., 2013). Now, personal monitoring section of BARC, Mumbai provides fast neutron personal monitoring service in India using CR-39 SSNTD polymer foils for industry and research institutions. CR-39 has got widespread acceptance for neutron monitoring in recent years (Tommasino et al., 1983). Lot of research has been carried out to study the use of CR-39 as neutron dosimeter. Recently, $CaSO_4$: Dy (Bhadane et al., 2016) and MgB_4O_7: Ce, Li (Yukihara et al., 2017) has been synthesized to study the neutron dosimetry that is explained as follows:

6.1.1. CR-39 as Neutron Dosimeter

CR-39 sheets were obtained from M/S Page Moulding Ltd. UK (Dhairyawan et al., 2003). The sheets were cut into small foils and were then exposed to various neutron energies from 0.45 to 4.4 MeV. The foils were then subjected to electrochemical etching at 60°C with AC electric high field of operating voltage 1250 V. This technique has made possible to monitor 100 keV neutrons (Dhairyawan et al., 2003). Later on the chemical etching parameters (etching time, temperature, normality of etchant etc.) for the use of CR-39 as a fast neutron dosimeter have been optimized (Castillo et al., 2013). CR-39 chips placed under 1.5 mm polyethylene radiator were exposed to different neutron fluencies. After several chemical etching processes with different conditions, the optimum characteristics for chemical etching were found at 6 N KOH solution, 60±1°C for 12 h. An accurate relation with a linear plot between the dose and track density was obtained. The linear curve plays an important role in determining the equivalent dose in the field of neutron dosimetry (kumar et al., 2010).

6.1.2. MgB_4O_7: Ce, Li Phosphor

MgB_4O_7 has already been described as a suitable material for TL dosimetry particularly using Mn and Dy dopants (Mckveer et al., 1995), due to its tissue equivalency, high sensitivity & interest for neutron dosimetry. The presence of enriched boron ^{10}B, due to high neutron capture cross section in the composition of host material opens the possibility of increased neutron sensitivity rather than ^{11}B. Moreover the dopant plays an important role in enhancing the TL sensitivity (Kano, 1999; Shi et al., 2008; Xie et. Al, 2010; Kang et al., 2010). MgB_4O_7: Ce, Li samples were synthesized by solid combustion synthesis (Yukihara et., 2017; Chick et al., 1990; Kingsley and Patil, 1988). Synthesized samples were irradiated by using uncalibrated PuBe source held in paraffin container (Yukihara et., 2017) and their TL properties were studied. The samples with enriched ^{10}B were found to show better TL intensity and hence MgB_4O_7: Ce, Li was found to be sensitive material to be used as neutron dosimeter.

6.1.3. $CaSO_4$ Microphosphor

Neutron dosimetry is complex as compared to gamma dosimetry because of the numerous direct & indirect energy transfer processes involved and the variation of reaction cross section with energy (Marrale et al., 2012; Douglas 1978). Although neutron has wide range of energy spectrum, the thermal neutrons are most important for neutron dosimetry.

Thermal neutron response of every TL material depends on the capture cross section of elements of TL materials. However $CaSO_4$ is reported on the basis of mixed field dosimetry (Ayyangar, 1974). Further $CaSO_4$: Dy microphosphor was synthesized using acid recrystallization method and the detailed synthesis is reported elsewhere (Bhadane et al., 2016). The microcrystalline thus obtained was irradiated using ^{252}Cf based thermal neutron source for fluence varying from $6x10^{11}$ n/cm^2 to $60x10^{11}$ n/cm^2 and hence their thermoluminescence properties were studied. The TL glow curve shows a single peak at 240°C, the intensity of glow peak increases with increase in neutron fluence without changing the peak position and shape. Further the TL response with respect to thermal neutron fluence was studied and it was observed that TL response increases linearly with increase in neutron fluence and beyond the fluence $52x10^{11}$ n/cm^2, TL response saturates. These results confirm that microphosphor could be promising for thermal neutron dosimetry and can also be used as routine TLD badge system in mixed field of gamma and neutrons for radiation workers.

CONCLUSION

Ionizing radiations are harmful to the human body. So, there is a need to measure even small radiation doses in the environment. High doses of ionizing radiation: gamma, electron, proton, neutrons, heavy ions exposure can result in skin diseases, hair loss, internal organ failure and death. Inspite of having high risk, radiations are used for diagnosis, treatment and research. Thermoluminescence (TL) is one of the important technique to estimate the quantity of absorbed dose. Today, there are a number of commercial TL dosimeters available for users. Still, scientific interest is prevailing worldwide with a view to develop good and efficient TL materials. Research is going on at a vast level and frequently newer and newer materials are being reported.

ACKNOWLEDGMENTS

Authors are grateful to Inter University Accelerator Centre (IUAC) New Delhi, for providing necessary facilities to carry out this work under research project (Project no. 61106 (BTR-I)).

REFERENCES

Akselrod, M.S.; Kortov, V.S.; Kravetsky, D.J.; Gotlib, V.I. *Rad., Prot. Dosim.* 1990, *32*, 15-22.

Ayyangar, K.; Lakshmanan, A.R.; Chandra, B.; Ramadas, K. *Phys. Med. Biol.* 1974, *19,* 665-668.

Aboelezz, E.; Sharaf, M.A.; Hassan, G.M.; El-Khodary, A.*J. of Lumin.* 2015, *166*, 156–161.

Amaldi, U.; Kra G. *Rep. Prog. Phys.* 2005, *68*, 1861–1882.

Bahl, S.; Lochab, S.P.; Kumar, P. *Radiat. Phys. Chem.* 2016, *119,* 136-141.

Bahl, S.; Lochab, S.P.; Pandey, A.; Aleynikov V.E.; Molokanov A.G.; Kumar, P. *Rad. Phys. Chem.* 2012, *81*, 1683–1687.

Barth, W.; Dahl, L.; Glatz, J.; Forck, P.; Klabunde, J. *Proceedings of the Particle Accelerator Conference*, CHICAGO, 2001; vol. 3281.

Barth, W.; Dahl, L.; Glatz, J.; Groening, L.; Richter, S.; Yaramishev, S. *Proceedings* DIPAC, 2003; vol. 161.

Bartlett, D.A.; Edwards, A.A. *Phys. Med. Biol.* 1979, *24,* 1276-83.

Bedyal, A.K.; Vinay, K.; Ntwaeaborwa, O.M.; Swart, H.C. *J. Alloys Comp.* 2017, *691,* 919-928.

Bedyal, A.K.; Vinay, K.; Swart, H.C. *J. Lumin.* 2017, *187*, 499–506.

Bhadane, M.S.; Dahiwale, S.S.; Sature K.R.; Patil, B.J.; Mandlik N.T.; Bhoraskar, V.N.; Dhole, S.D. *J. Alloys Comp.* 2017, *695*, 1918-1923.

Bhadane, M.S.; Mandlik, N.K.; Patil, B.J.; Dahiwale, S.S., Sature, K.R., Bhoraskar, V.N., Dhole, S.D. *J. Lumin.* 2016, *170,* 226–230.

Binder W.; Disterhoft, S.; Cameron, J.R. *In: Proc. 2nd Conf. on Luminescence Dosimetry,* Gatlinburg. Conf; 680920; NTIS, Springfield, VA, 1968; p. 113.

Bilski, P.; Budzanowski, M.; Hoffmann, W.; Molokanov, A.; Olko, P.; Waligorski, M.P.R. *Rad. Prot. Dosim.* 1997, *70*, 501-504.

Bilski, P.; Budzanowski, M.; Olko, P.; Waligorski, M.P.R. *Radiat. Meas.* 1998, *29*, 355-362.

Bilski, P.; Olko, P.; Burgkhardt, B.; Piesch, E.; Waligorski, M.P.R. *Radiat. Prot. Dosim* 1994, *55*, 31-34.

Brahme, A.; Svensson, H. *Acta Radiol. Oncol.* 1979, *18*, 244-250.

Brahme, A.; Hulten, G.; Svensson, H. *Phys. Med. Bio.* 1975, *20*, 39-45.

Briand, J.P.; de Billy, L.; Charles, P., Essabaa, S., Briand, P., Geller, R., Desclaux, J.P., Bliman, S.; Ristori, C. *Phys. Rev. Lett.* 1990, *65*, 159-164.

Burkert, S.; Kuntzsch, M.; Bellmann, C.; Uhlmann P.; Stamm, M. *Appl. Surf. Sci.* 2009, *255*, 6256-6260.

Cassoll, R.M.; Benton, E.V. *Nuc. Track detec*. 1978, *2,* 173-179.

Castillo, F.; Espinosa, G.; Golzarri, J.I.; Osorio, D.; Rangel, J.; Reyes, P.G.; Herrera, J.J.E. *Radiat. Meas.* 2013, *50,* 71-73.

Chen, R.; Fogel, G. *Radiat. Prot. Dosim.* 1993, *47*, 23-26.

Chithambo, M.L.; Wako, A.H.; Finch, A.A. *Radiat. Meas.* 2017, *97*, 1-13.

Chick, L.A.; Pederson, L.R.; Maupin, G.D.; Bates, J.L.; Thomas, L.E.; Exarhos, G.J. *Mater. Lett.* 1990, *10*, 6–12.

Chopra, V.; Singh, L.; Lochab, S.P.; Aleynikov, V.E.; Oinam, A.S. *Radiat. Phys. Chem.* 2014, *102*, 5-10.

Chopra, V.; Singh, L.; Lochab, S.P. *Nucl. Instrum. Meth. Phys. Res. Section A*. 2013, *717*, 63-68.

Chopra, V.; Singh, P.; Singh L. *Rad. Phys. Chem.* 2012, *81,* 771-75.

Dabrel, K.V.; Dhoble, S.J. *Adv. Mat. Lett.* 2013, *4*, 921-926.

Dalal, M.; Taxak, V.B.; Chahar, S.; Khatkar, A.; Khatkar S.P. *J. Phy. Chem. Solids.* 2016, *89*, 45-52.

Daniels, F.; Boyd, C.A.; Saunders, D.F. *Science.* 1953, *117,* 343-49.

Del. Nery, S.M.; Maigon, P.W.; Isotoni, S.; Graham, R. *Phys. Rev. B.* 1994, *49,* 3760-3764.

De Almeida, C.E.; Almond, P.R. *Radiology*. 1974 a, *111*, 439-445.

Devicienti, S.; Strigari, L.; D'Andrea, M.; Benassi, M.; Dimiccoli, V.; Portaluri, M. *J. Exp. Clin. Cancer Res.*, 2010, *13*, 29–47.

Dhairyawan, M.P.; Marathe, P.K.; Massand, O.P. *Radiat. Meas.* 2003, *36*, 435- 438.

Douglas, J.A.; *The Applications of TL Materials in Neutron Dosimetry*; AERE-R-9196 UKAEA Atomic Energy Research Establishment, Harwell, Rep. Adam Hilger, Bristol, 1978; 229.

Dutta, S.; Som, S.; Kunti, A.K.; Kumar, V.; Sharma, S.K.; Swart, H.C.;Visser, H.G. *Acta Mater.* 2017,124, 109–119.

Efstathopoulos, E.P.; Makrygiannis, S.S.; Kottou, S.; Karvouni, E.; Giazitzoglou, E.; Korovesis, S.; Tzanalaridou, E.; Raptou, P.D.; Katritsis, D.G. *Phys. Med. Bio*. 2003, *48*, 3059-3068.

Eickhoff, H.; Haberer, T.; Kra, G.; Krause, U.; Richter, M.; Steiner R.; Debus, J.; Strahlenther. *Oncology*. 1999, *175,* 21-25.

Fox, P.J.; Akber, R.A.; Prescott, J.R.J. *Phys. D. Appl. Phys.* 1988, *21,* 189-193.

Furtado, W.W.; Tome, T.; Isotani, S.; Antonini, A.R.; Black, A.R.; Pontuschka, W.M.; Rabbani, S.R. *Anais da Academia Brasileiria de ciencias* 1989, *61,* 397-403.

Gai, M.Q.; Chen, Z.Y.; Fan, Y.W.; Yan, S.Y.; Xie, Y.X.; Wang, J.H.; Zhang, Y.G. *Radiat. Meas.* 2015, *78,* 48-52.

Gupta, K.K.; Kadam, R.M.; Dhoble, N.S.; Lochab, S.P.; Singh, V.; Dhoble S.J. *J. Alloys Comp.* 2016, *688*, 982- 993.

Gooden, D.S.; Brickner, T.J. *Radiology* 1972, *102*, 685–689.

Haberer, T. *AIP Conf. Proc.* 2002, *610,* 157-162.

Handee, W.R.; Ibott, G.S.; Gilbert, D.B. *Inter. J. Applied Rad. Isot.* 1968, *19,* 431-437.

Hankins, D. E. *A small inexpensive albedo neutron dosimeter.* UCLA report No. 5261, UC-41, 1973.

Jayasudha, S.; Madhukumar, K.; Nair, C.M.K., Nair, R.G.; kumar, A.; Elias T.S. *Spect. Acta Part A: Molec. Biomol. Spectr.* 2016, *155*, 21–27.

Kano, T. Principal phosphor materials and their optical properties, *in: Phosphor Handbook*, CRC, Boca Raton, 1999, 177–200.

Kang, M.; Yan, W.; Liu, J.; Sun, R.; Wuhan, J. *Univ. Technol.-Mater. Sci. Ed.* 2010, *25,* 179–183.

Kazakis, N.A.; Tsetine, A.T.; Kitis, G.; Tsirliganis N.C. *Radiat. Meas.* 2016, 89, 74-81.

Kore, B.P.; Dhoble, N.S.; Kadam, R.M.; Lochab, S.P.; Singh, M.N.; Dhoble, S.J.; Swart, H.C. *J. Phys. D. Appl. Phys.* 2016, *49*, 95102-113.

Kingsley, J.J.; Patil, K.C. *Mater. Lett.* 1988, *6*, 427–432.

Kitis, G.; Polymeris, G.S.; Sfampa, I.K.; Prokic, M.; Meriç, N.; Pagonis, *V. Rad. Meas.* 2016, *84*, 15-25.

Knoll, G.F. *Radiation detection and measurement.* 3rd edition. John Wiley & Sons Inc., NEW YORK. 1999.

Kron, T. *Radiat. Prot. Dosim.* 1999, *85*, 333-340.

Kore, B.P.; Dhoble, N.S.; Kadam, R.M.; Lochab, S.P.; Dhoble, S.J. *Mater. Chem. Phys.* 2015, 161, 96-106.

Kore, B.P.; Dhoble, N.S.; Lochab, S.P.; Dhoble, S.J. *RSC Adv.*, 2014, *4,* 49979-985.

Kore, B.P.; Dhoble, N.S.; Lochab, S.P.; Dhoble, S. J. *J. Lumin.* 2014, *145,* 299–306.

Kulig D.W.; Gieszczyk, W.; Bilski W.P.; Marczewska, B.; Kosowski, M. *Rad. Meas.* 2016, *85*, 88-92.

Kumar, V.; Sonkawade, R.G.; Dhaliwal, A.S. *Indian J. Pure Appl. Phys.* 2010, *48,* 466-469.

Lakshmanan, A.R. *Nuc. Tracks* 1982, *6,* 59-78.

Larsson L.; Katz, R. *Nucl. Instr. Methods.* 1976, *138*, 631-636.

Lawrence, T.S.; Feng, M. *J. Natl. Cancer Inst.* 2013*, 105,* 7–8.

Leo, W.R. *Techniques for nuclear and particle physics experiments-A how to approach. 2nd* revised edition. Springer-Verlag, BERLIN; 1994.

Lochab, S.P.; Salah, N.; Sahare, P.D.; Chauhan, R.S.; Ranjan, R. *Nucl. Instrum. Methods B* 2007, *254,* 231-235.

Lochab, S.P.; Kanjilal, D.; Salah, N.; Habib, S.S.; Lochab, J.; Ranjan, R.; Aleynikov, V.E.; Rupasov, A.A.; Pandey, A. *J. Appl. Phys.* 2008, *104,* 033520-525.

Lochab, S.P.; Sahare, P.D.; Chauhan, R.S.; Salah, N.; Pandey, A. *J. Phys. D* 2007,*40*, 1343-1348.

Lochab, S.P.; Pandey, A.; Sahare, P.D.; Chauhan, R.S.; Salah, N.; Ranjan, R. *Phys. Stat. Sol. (a)* 2007, 204, 2416-425.

Lucas, A.C.; Kaspar, B. In: *Proc 5th Conf. on Luminescence Dosimetry,* Sau Paulo. I Phys. Inst. Univ. GIESSEN; 1977, pp-131.

Mandlik, N.K.; Sahare, P.D.; Kulkarni, M.S.; Bhatt, B.C.; Bhoraskar, V.N.; Dhole, S.D. *J. Lumin.* 2014, 146, 128–132.

Mandowska, E.; Bilski, P.; Ochab, E.; Wiatek, J.; Mandowski, A. *Radiat. Prot. Dosim.* 2002, *100,* 451–454.

Marcazzó, J.; Santiago, M.; D'Angelo C.; Furetta C.; Caselli, E. *Nucl. Instrum. Methods Phys. Res. B* 2010, *268,* 183–186.

Marrale M.; Longo, A.; Bartolotta, A.; Basile, S.; D'Oca, M.C.; Tomarchio, E.; Cirrone, G.A.P.; Di Rosa, F.; Romano, F.; Cuttone, G.; Brai, M. *Nucl. Instrum. Methods B* 2012,) *292*, 55-60.

Mathur,V.K.; Lewandowski, A.C.; Guardala, N.A.; Price, J.L. *Radiat. Measur.* 1999, 30, 735-738.

McKeever, S.W.S.; Moscovitch, M.; Townsend, P.D. *Thermoluminescence Dosimetry Materials: properties and Uses,* Nuclear Technology Publishing, Ashford, 1995.

Mesterhzy D.; Osvay, M.; Kovcs, A.; Kelemen, A. *Radiat. Phys. Chem.* 2012, *81,* 1525-1527.

Mehta, S.K.; Sengupta, S. *Health Phys.* 1976, *31*, 176-180.

Nakajima, T.; Murayama, Y.; Matsuzawa, T., Koyano, A. *Nucl. Instum. Methods* 1978, *157,*155-160.

Naik, R.; Prashantha, S.C.; Nagabhushana, H.; Sharma, S.C.; Nagaswarupa, H.P.; Girish, K.M. *J. Alloys Compd.* 2016, *682*, 815-824.

Newhauser, W.D.; Zhang, R. *Phys. Med. Biol.*2015, *60,* R155-R 209.

Noh, A.M.; Amin, Y.M.; Mahat, R.H.; Bradley D.A. *Radiat. Phys. Chem.* 2001, *61,* 497-499.

Nunes, M.G.; Campos, L.L. *Rad. Measur.* 2008, *43,* 459-462.

Oliveira, L.C.; Yukihara, E.G.; Baffa, O. *Sci. Rep.* 2016, *6*, 24348-355.

Oza, A.H.; Dhoble, N.S.; Lochab, S.P.; Dhoble, S.J. *Lumin: J. Biol. Chem. Lumin.* 2015, *30,* 967-77.

Pandey, A.; Sharma, V.K.; Mohan, D.; Kale, R.K.; Sahare, P.D. *J. Phys.* 2002a, *35*, 1330–1333.

Pandey, A.; Sahare, P.D.; Shahnawaz; Kanjilal, D *J. Phys. D: Appl. Phys.* 2004, *37,* 842-46.

Pandey, A.; Sonkawade, R.G.; Sahare, P.D. *J. Phys. D: Appl. Phys.* 2002, *35*, 2744-750.

Patterson, D.A. *Rev. Sci. Instrum.* 1962, *33*, 831–833.

Prokic, M. *Nucl. Instum. Methods* 1980, *175*, 83-90.

Podgorsak E.B., *Radiation Physics for Medical Physicists*, Springer-Verlag, Berlin, Heidelberg; 2006, pp. 10–14.

Rani, R.S.; Lakshmanan, A. *J. Lumin.* 2016*, 174*, 63–69.

Rawat, N.S.; Dhabekar, B.; Muthe, K.P.; Koul, D.K.; Datta, D. *Nucl. Instr. Meth. in Phys. Res. B,* 2017, *397*, 27–32.

Ribeiroa, D.R S.; Souzaa*,* D.N.; Maiaa, A.F.; Baldochib, S.L.; Caldas, L.V.E. *Radiat. Meas*. 2008, *43*, 1132 – 1134.

Roman-Lopeza J., Lozanob, I.B.; Cruz-Zaragozac, E.; Guzman Castañedad, J.I.; Díaz-Góngora, J.A.I. *App. Radiat. and Isot.* 2017, *124,* 44–48.

Rudén, B. *Acta Radiol. Ther. Phys. Biol.* 1976, *15,* 447–464.

Sadel, M.; Bilski, P.; Swako, J.; Rydygier, M.; Horwacik T.; Weber A. *Rad. Meas.* 2015, *82*, 8-13.

Sahare, P.D.; Moharil, S.V. *J. Phys. D: Appl. Phys.* 1990, *23,* 567-570.

Sahare, P.D.; Singh, M.; Kumar, P. *J. Lumin.* 2015, *160*, 158–164.

Sahare, P.D.; Salah, N.; Lochab, S.P.; Mohanty, T.; Kanjilal, D. *J. Phys. D: Appl. Phys.* 2005, *38,* 3995-3999.

Sahare, P.D.; Ranjan, R.; Salah, N.; Lochab, S.P. *J. Phys. D* 2007, *40*, 759-764.

Salah, N.; Habib, S.; Babkair, S.S.; Lochab, S.P.; Chopra, V. *Radiat. Phys. Chem*. 2013, *86*, 52-58.

Salah, N.; Najlaa, D.A.; Habib, S.S.; Lochab, S.P.; *J. Lumin.* 2015, *167,* 59–64.

Salah, N.; Najlaa, D.A., Habib, S.S., Lochab, S.P. *Nucl. Instrum. Methods Phys. Res. B* 2015, *358*, 201–205.

Salah, N.; Sahare, P.D.; Lochab, S.P.; Kumar, P. *Radiat. Meas.* 2006, *41*, 40-47.

Salah, N.; Sahare, P.D.; Rupasov, A.A. *J. Lumin.* 2006, *124,* 357-364.

Salah, N.; Habib, S.S.; Khan Z.H.; Al-Hamedi, S.; Lochab, S.P. *J. Lumin.* 2009, *129,* 192–196.

Salah, N.; Lochab, S.P.; Kanjilal, D.; Ranjan, R.; Rupasov, A.A.; Khan, Z.H.; Habib, S.S.; Aleynikov, V.E. *IEEE Trans. Nanotechnol* 2008, *7,* 749-755.

Salah, N.; *Rad. Phys. Chem.* 2011, *80,* 1–10.

Salah, N.; Sahare, P.D.; Prasad, A. *J. Lumin.* 2006, *121,* 497-503.

Salah, N. *J. Phys. D: Appl. Phys.* 2008, *41,* 155302-306.

Salah, N.; Lochab, S.P.; Ranjan, R.; Kanjilal, D.; Habib, S.S.; Aleynikov, V.E.; Rupasov, A.A. *J. Appl. Phys.* 2007, *102*, 64904-910.

Salah, N., Sahare, P.D. *Rad. Meas.* 2006, *41*, 665 – 670.

Salah, N.; Sahare, P.D.; Shahnawaz; Lochab, S.P. *Radiat. Eff. Def. Solids* 2004, *159*, 321–334.

Schulman, J.H.; Kirk, R.D.; West, E.J. In: *Proc. 1st Int. Conf. on Luminescence Dosimetry*, Standford, 1965, pp. 113-118.

Schardt, D.; Elsässer, T.D.; Schulz-ertner. *Rev. Mod. Phys.* 2010, *82,* 383–425.

Shinde, S.S.; Dhabekar, B.S.; Gundu Rao, T.K.; Bhatt, B.C. *J. Phys. D* 2001, *34,* 2683-688.

Shinde, N.; Dhoble, N.S.; Gedam, S.C.; Dhoble. S.J. *Radiat. Eff. Def. Solids* 2014, *169*, 361-367.

Shi, S.; Gao, J.; Zhou, J. *Opt. Mater.* 2008, *30*, 1616–1620.

Singh, L.; Chopra,V.; Lochab, S.P. *J. Lumin.* 2011,*131*, 1177–1183.

Singh, L.; Chopra, V. *J. Mat. Sci. Tech.* 2011, *27*, 967-72.

Soares, A.F.; Tatumi, S.H.; Mazzo, T.M.; Rocca, R.R. Courrol, L.C. *J. Lumin.* 2017, *186,* 135–143.

Som, S.; Sharma, S.K.; Lochab, S.P.; Swift, I. *Phys. Status Solidi (a)* 2013, *210,* 1624–1635.

Stochioiu, A.; Georgescu, I. *Rom. Rep. Phys.* 2003, *55,* 136–139.

Suntharalingam, N.; Cameron, J.R. *Phys. Med. Biol.* 1969, *14,* 397-410.

Sunta, C.M.; Yoshimura, E.M.; Okuno, E. *J. Phys. D. Appl. Phys.* 1994, *27,* 852-860.

Takenga, M.; Yamamoto, O.; Yamashita, T. *Nucl. Instum. Methods* 1980, *175*, 77-80.

Tochilin E., Goldstein N.; Miller, W.S. *Health Phys.* 1969, *16,* 1-5.

Tommasino, L.; Zapparoli, G.; Spiezia, P.; Grilth, R.V.; Espinosa; Gracia, G., Berent, Di. *Proceedings of the 12th International Conference on SSNTD,* Acapulco, Mexico, 1983.

Tiwari G.; Brahme, N.; Sharma, R.; Bisen, D.P.; Sao, S.K.; Khare. A. *J. Lumin.* 2017, *183,* 89–96.

Tamrakar R.K.; Bisen, D.P.; Upadhyay K. *Radiat. Phys. Chem.* 2017, *130,* 321–334.

Vallejo, M.Á.; Modesto, A.S.; María, L.V.; Juan, C.A.; Ricardo, N.; Emma, E.; Luis, A.D. *J. Lumin.* 2017, *182*,160–165.

Veronese, I.; Galli, A.; Cantone, M.C.; Martini, M.; Vernizzi, F.; Guzzi, G. *Radiat. Meas.* 2010, *45,* 35-41.

Venkateswara, R.G., Yadagiri, R.P.; Veeraiah, N. *Mater. Letter.* 2002, *57,* 403-08.

Wani, J.A.; Dhoble, N.S.; Lochab, S.P.; Dhoble. S.J. *Nucl. Instrum. Meth. Phys. Res. B* 2015, *349*, 56–63.

Wiedmann, E.; Schmidt, G. *Ann. Rev. Phys. Chem.* 1895, *54,* 604-25.

Wilson R.R. *Radiology*, 1946, *47,* 487–491.

Xie, A.; Yuan, X.; Wang, F.; Shi, Y.; Mu Z. *J. Phys. D: Appl. Phys*. 2010, *43,* 055101-106.

Yang, B., Xie, L., Xu, Y., Townsend, P.D., 2002. *Nucl. Instrum. Methods B* 187, 408.

Yerpude, M.M.; Dhoble, N.S.; Lochab, S.P.; Dhoble, S.J. *J. Lumin.* 2016, *31,*115-124.

Yukihara, E.G.; Doull, B.A.; Gustafson T.; Oliveira L.C.; Kurt, K.; Milliken, E.D. *J. Lumin.* 2017,*183*, 525–532.

Zahedifar, M.F.; Almasifard, E.; Sadeghi, M.; Kashefi B.; Ramazani A.; Moghaddam, A. *Radiat. Phys. Chem.* 2016, *125*, 127–133.

Zhao, Y.; Wang, Y.; Cui, M.; Jin, H.; Fan, Z.; Zhao, C.; Wu, X.; Townsend P.D. *Radiat. Meas.* 2017, *99*, 44-49.

Biographical Sketches

Vibha Chopra, PhD

Affiliation: P.G. Department of Physics & Electronics, DAV College, Amritsar - 143001, Punjab, India

Education: MSc, PhD, Physics

Research and Professional Experience: 10 Years

Professional Appointments: Working as an Assistance Professor

Honors: Young Scientist Award by Venus International Foundation, Chennai, India

Publications from the Last 3 Years:

List of Publications

Research papers published in Journals

1. Size and shape-tailored hydrothermal synthesis and characterization of nanocrystalline $LaPO_4:Eu^{3+}$ phosphor, Sudeshna Ray, Govind B. Nair, Prachi Tadge, Nidhi Malvia, Vinita Rajput, Vibha Chopra and S.J. Dhoble, *Journal of Luminescence*, 194(2018)64-71.
2. Impact of Ga^{3+} on the phase transition and luminescence properties of substituted $Sr_2SiO_4:Eu^{2+}$ phosphor synthesized from a highly

homogeneous oxide precursor based on a novel water-soluble silicon compound. Sudeshna Ray, Prachi Tadge, Govind B. Nair, Vibha Chopra, S.J. Dhoble. *Ceramics International*, 44 (2018) 5506-5512.
3. Impact of C^{5+}ion beam on Dy activated $Sr_2B_5O_9Cl$ TL phosphor. Abha H.Oza, Vibha Chopra, N.S. Dhoble and S.J. Dhoble. *J. of Material science and electronics*. (2018) Accepted.
4. Investigation of optical properties of $70B_2O_3$-$29Bi_2O_3$-iDy_2O_3)XBT glasses under the influence of gamma irradiation, Vanita thakur, Anupinder singh, Vibha Chopra, Sonika Thakur and Lakhwant Singh, *Current reports on science & technology*, 02(2016)100-104.
5. TL dosimetry of nanocrystalline $Li_2B_4O_7$: Cu irradiated to 120 MeV Ag^{9+} ion beam, Vibha Chopra, Lakhwant Singh and S.P Lochab, *Current reports on science & technology*,01(2015)1-4.
6. TL dosimetry of nanocrystalline $Li_2B_4O_7$:Cu exposed to 150 MeV proton, 4 MeV and 9 MeV electron beam, Vibha Chopra, Lakhwant Singh, S.P. Lochab, V.E. Aleynikov, Arun S. Oinam, *Radiation Physics and Chemistry*,102(2014), 5-10.

Prof. S. J. Dhoble, PhD

Affiliation: Department of Physics, R.T.M. Nagpur University, Nagpur-440033, India

Education: MSc, PhD, Physics

Business Address: Department of Physics, R.T.M. Nagpur University, Nagpur-440033, India

Research and Professional Experience: 24 Years

Professional Appointments: Working as a Professor

Honors: Fellow, Luminescence Society of India

Publications from the Last 3 Years:

Research papers published in Journals

1. Impact of Ga^{3+} on the phase transition and luminescence properties of substituted $Sr_2SiO_4:Eu^{2+}$ phosphor synthesized from a highly homogeneous oxide precursor based on a novel water-soluble silicon compound. Sudeshna Ray, Prachi Tadge, Govind B. Nair, Vibha Chopra, S.J. Dhoble. *Ceramics International*, 44 (2018) 5506-5512.
2. Impact of C^{5+}ion beam on Dy activated $Sr_2B_5O_9Cl$ TL phosphor. Abha H.Oza, Vibha Chopra, N.S. Dhoble and S.J. Dhoble. *J. of Material science and electronics*. (2018) Accepted.
3. Microstructural evolution and photoluminescence performanance of nickel and chromium doped ZnO nanostructures, J.S. Tawale, Ashavani Kumar, G. Swati, D. Haranath, S.J. Dhoble and A.K. Srivastava, *Materials Chemistry and Physics*, 205(2018)9-15.
4. Structural and optical characterization of RE (Eu^{2+}, Ce^{3+}) doped $SrMg_2Al_6Si_9O_{30}$ nanocrystalline phosphor, M.M. Kolte, V.B. Pawade and S.J. Dhoble, *Journal of Materials Science: Materials in Electronics*, 29 (2018) 906–913.
5. *Recent advances and developments on integrating nanotechnology with chemiluminescence assays*, Ashish Tiwari and S.J. Dhoble, Talanta, 180 (2018) 1-11.
6. On the study of C^{6+} ion beam and γ-ray induced effect on structural and luminescence properties of Eu doped $LiNaSO_4$:Explanation of TSL mechanism by using PL, TL and EPR study, Karan Kumar Gupta, Ramakant Kadam, N.S. Dhoble, S P Lochab and S J Dhoble, *Physical Chemistry Chemical Physics*, 20(2018) 1540-1559.
7. Energy transfer mechanisms and optical thermometry of $BaMgF_4:Yb^{3+},Er^{3+}$ phosphor, Bhushan Kore, Ashwini Kumar, Lucas Erasmus, Robin Kroon, Jacobus Terblans, Sanjay Dhoble and Hendrik Swart, *Inorganic Chemistry*, 57(2018)288-299.
8. Study of lyoluminescence of gamma-irradiated $GdCa_4O(BO_3)_3$:Ce phosphors: mass of solute and pH of solvent as a key factor, G.C.

Mishra, S.J. Dhoble, Ashok Srivastava, Surendra Dwiwedi and R.S. Kher, *Optik*,158(2018)826-830.

9. Luminescence in microcrystalline green emitting $Li_2Mg_{1\_x}ZrO_4$:xTb^{3+}(0.1 < x < 2.0) phosphor, V.R. Panse, N.S. Kokode, K.N. Shinde and S.J. Dhoble, *Results in Physics*, 8 (2018) 99–103.
10. Optical analysis of RE^{3+} (RE=Eu, Tb): CdO-PbO-B_2O_3-AlF_3 glasses, B. Suneethaa, S. Sailaja, S.J. Dhoble and B. Sudhakar Reddyd, *Optik*,155 (2018) 216–224.
11. Size and shape-tailored hydrothermal synthesis and characterization of nanocrystalline $LaPO_4$:Eu^{3+} phosphor, Sudeshna Ray, Govind B. Nair, Prachi Tadge, Nidhi Malvia, Vinita Rajput, Vibha Chopra and S.J. Dhoble, *Journal of Luminescence*, 194(2018)64-71.
12. Novel photoluminescence properties of Eu^{3+} doped chlorapatite phosphor synthesized via sol-gel method, Arti Durugkar, Sumedha Tamboli, N.S. Dhoble and S.J. Dhoble, *Materials Research Bulletin*, 97 (2018) 466-472.
13. Efficient red downshifting in layered structure: A broad spectral converter for enhancing photo-response of solar cell, Sonal P. Ghawade, Kavita A. Deshmukh, D.R. Peshwe, S.J. Dhoble and Abhay D. Deshmukh, *Journal of Inorganic and organic and Organometallic Polymers and Materials*, 27(1) (2017) 173-180.
14. Synthesis and Optical Characterization of $CaSr_2Al_2O_6$:$RE^{3+($RE=Eu and Tb) Phosphor for solid state lighting,, Vishal R. Panse, A.N. Yerpude, N.S. Kokode, S.J. Dhoble and Renu Choithrani, *Journal of Materials Science: Materials in Electronics*, 28(22)(2017)16880-16887.
15. High dose TL response of fly ash collected from coal fired thermal power plant, Swati Joshi, Karan Kumar Gupta, S.K. Ubale and S.J. Dhoble, *Radiation Measurements*,105(2017)47-53.
16. Optical and bandgap study of rare earth doped phosphate phosphor,V.B. Pawade, A. Zanwar, R.P. Birmod, S.J. Dhoble and L.F. Koao, *Journal of Materials Science: Materials in Electronics*, 28(21)(2017)16303-10313.

17. Structural properties of CuxNi1-xFe2O4 nano ferrites prepared by urea-gel microwave auto combustion method, H.S. Ahamad, N.S. Meshram, S.B. Bankar, S.J. Dhoble and K.G. Rewatkar, *Ferroelectrics*, 516 (2017) 67–73.
18. White light emission through efficient energy transfer from Ce^{3+} to Dy^{3+} ions in $Ca_3Mg_3(PO_4)_4$ matrix aided by Li^+ charge compensator, Govind B. Nair and S.J. Dhoble, *Journal of Luminescence*, 192(2017)1157-1166.
19. Analysis of growth kinetics and impact of NH_3 on the morphology evolution of hexagonal-prism shaped $Y_4O(OH)_9NO_3/Y_2O_3$ single crystals, Sudeshna Ray, Govind B. Nair, S.J. Dhoble, H. Kato, Masato Kakihana and Teng Ming Chen, *Materials Research Bulletin*, 95 (2017) 597-606.
20. Deep blue light emitting Cyno-DPQ phosphor with Large Stokes Shift and High Thermal Stability for OLEDs and display applications, Minakshi Ghate, H.K. Dahule and S.J. Dhoble, *Optik*, (2017)198-205.
21. Structural and optical investigation of novel europium hybrid organic complex for OLEDs and solid state lighting, Dipti Chitnis, N. Thejo Kalyani, S.J. Dhoble, *Journal of Materials Physics and Chemistry*, 5(1) (2017) 1-10.
22. TL glow curve analysis and kinetics of UV, β and γ irradiated $YBO_3:Eu^{3+}$ and Y_2O_3: Eu^{3+} phosphors, Vikas Dubey, Neha Dubey, S.J. Dhoble and Hendrik C. Swart, *Journal of Materials Science: Materials in Electronics*, 28 (2017) 13565–13578.
23. Mechanical and UV absorption behavior of zinc oxide nanoparticles: reinforced poly(vinyl alcohol-g-acrylonitrile) nanocomposite films, S. Chouhan, A.K. Bajpai, J. Bajpai, R. Katare, S.J. Dhoble, *Polymer Bulletin*, 74(10) (2017) 4119-4141.
24. Photoluminescence and thermoluminescence of $K_2Mg(SO_4)_2$:Eu and evaluation of its kinetic parameters, Archana Deshpande, N.S. Dhoble, S.C. Gedam and S.J. Dhoble, *Luminescence: The Journal of Biological and Chemical Luminescence*, 32(5), (2017) 757-764.

25. Optical characterization of zinc lithium bismuth borate glasses doped with Tb^{3+} for novel applications, K. Mallikarjuna, M. Bhushana Reddy, G. Moulika, B. Naveen Kumar Reddy, R. Ramanaiah, S.J. Dhoble and B. Sudhakar, *AIP Conference Proceedings* 1832(1) (2017) 070004-1 -070004-3.
26. VUV-UV photoluminescence properties of Ce^{3+} doped Ca_2SnO_4 phosphor for PDP application, M. Srinivas, Sumedha Tamboli and S.J. Dhoble, *Optik*, 145 (2017)202-208.
27. Luminescence and structural properties of Gd_2SiO_5:Eu^{3+} phosphors synthesized from the modified solid state method, Yogita Parganiha, Jagjeet Kaur, Neha Dubey, Vikas Dubey, Ravi Shrivastava, S.J. Dhoble and Hendrik C. Swart, *Ceramics International*, 43 (2017) 9084–9091.
28. Mechanoluminescence characterization of europium doped $Ca_3Si_2O_4N_2$ phosphor, A.K. Sahu, Renu Nayar, V. Nayar and S.J. Dhoble, *Bionano Frontier*, (2017) 241-243.
29. Luminescence of Ce^{3+} ion in $NaSr_4(BO_3)_3$ blue emitting phosphor, Sumedha Tamboli, M. Srinivasa and S.J. Dhoble, *Bionano Frontier*, (2017) 261-262.
30. VUV-UV photoluminescence properties of Eu^{3+} doped Ca_2SnO_4 phosphor, Sumedha Tamboli, M. Srinivasa and S.J. Dhoble, *Bionano Frontier*, (2017) 263-266.
31. Synthesis and photoluminescence properties of red, green and blue emitting $LaYO_3$:M (M = Eu^{3+}, Tb^{3+}, Sm^{3+},Bi^{3+}, Pb^{2+}) phosphors, D.R. Taikar, Sumedha Tamboli and S.J. Dhoble, *Optik*, 142 (2017) 183–190.
32. Thermoluminescence and kinetic study of low z $LiAlO_2$:Dy/Eu phosphors, Monali R. Kadukar, P.W. Yawalkar, S.P. Lochab, S.J. Dhoble, and B. Sudhakar Reddy, *Ferroelectrics Letters* Section, 44(2017)8-17.
33. Phosphor for phototherapy: Review on psoriasis, P.S. Hemne, R.G. Kunghatkar, S.J. Dhoble, S.V. Moharil and V. Singh, *Luminescence: The Journal of Biological* and *Chemical Luminescence*, 32 (2017)260-270.

34. Theoretical analysis of electron vibration interactions and study of photo physical properties in Ce^{3+} doped $Ca_2P_2O_2$ nano phosphor capped with SHMP, Abhilasha Jain, D.R. Peshwe, Ashwini Kumar, S.J. Dhoble, M.M. Yerpude, Govind B. Nair, H.C. Swart, *Materials Chemistry and Physics*, 196 (2017) 213-221.
35. Influence of Li^+ charge compensator ion on the energy transfer from Pr^{3+} to Gd^{3+} ions in $Ca_9Mg(PO_4)_6F_2:Gd^{3+}$, Pr^{3+}, Li^+ phosphor, Sumedha Tamboli and S.J. Dhoble, *Spectrochimica Acta Part A: Molecular and Biomolecular Spectroscopy*, 184 (2017) 119–127.
36. Critical analysis of phase evolution, morphological control, growth mechanism and photophysical applications of ZnS nanostructures (0D-3D): A review, Ashish Tiwari and S.J. Dhoble, *Crystal Growth and Design*, 17(1)(2017)381-407.
37. Spectroscopic investigation of up-conversion properties in green emitting $BaMgF_4:Yb^{3+},Tb^{3+}$ phosphor, Bhushan P. Kore, Ashwini Kumar, Anurag Pandey, Robin E. Kroon, Jacobus J. Terblans,Sanjay J. Dhoble and Hendrik C. Swart, *Inorg. Chem.*, 56(2017) 4996–5005.
38. Investigation of luminescence processes in $Li_6Gd(BO_3)_3:Eu^{3+}$ phosphor, Mrunal M. Yawalkar, G.D. Zade, Vijay Singh and S.J. Dhoble, *J Mater Sci: Mater Electron*, 28(1)(2017)180-189.
39. Studies of radiation-induced defects in Li_2SiO_3:Sm phosphor material, N. Singh, Vijay Singh, S. Watanabe, T.K. Gundurao, J.F.D. Chubaci, N.F. Cano, M.S. Pathak, Pramod K. Singh and S.J. Dhoble, *Journal of Electronic Materials*, 46(1) (2017)451-457.
40. White light emitting $MZr_4(PO_4)_6:Dy^{3+}$ (M = Ca, Sr, Ba) phosphors for WLEDs, Govind B. Nair and S.J. Dhoble, *J Fluoresc.*, 27(2)575-585.
41. Luminescence analysis of $SrGa_2Si_2O_8$: RE^{3+} (RE= Dy, Tm) phosphors, Monali R. Kadukar, S.J. Dhoble, A.K. Sahu, V. Nayar, S. Sailaja, B. Sudhakar Reddy, *Luminescence: The Journal of Biological and Chemical Luminescence*, 32(2)(2017)159-170.
42. Dy^{3+}, Sm^{3+}, Ce^{3+} and Tb^{3+} activated optical properties of microcrystalline $BaMgP_2O_7$ phosphors, J.A. Wani, N.S. Dhoble,

N.S. Kokode, Vijay Singh and S.J. Dhoble, *Luminescence: The Journal of Biological and Chemical Luminescence*, 32(2017)240-252.

43. Exploration of electron-vibrational interaction in the 5d states of Eu^{2+} ions in $ABaPO_4$ (A = Li, Na, K and Rb) phosphors, Govind B. Nair, Priyanka D. Bhoyar and S.J. Dhoble, *Luminescence: The Journal of Biological and Chemical Luminescence*, 32(1)(2017)22-29.
44. Orange light emitting $Ca_3Mg_3(PO_4)_4:Sm^{3+}$ phosphors, Govind B. Nair and S.J. Dhoble, Luminescence: *The Journal of Biological and Chemical Luminescence*, 32(1)(2017)125-128.
45. Resonant and non-resonant energy transfer from $Ce^{3+} \rightarrow X$ ($X=Tb^{3+}$, Eu^{3+} or Dy^{3+}) in $NaMgSO_4F$ material, S.R. Choubey, S.C. Gedam and S.J. Dhoble, *Luminescence: The Journal of Biological and Chemical Luminescence*, 32(2)(2017)253-256.
46. A comparative investigation of Ce^{3+}/Dy^{3+} and Eu^{2+} doped $LiAlO_2$ phosphors for high dose radiation dosimetry: Explanation of defect recombination mechanism using PL, TL and EPR study, Karan Kumar Gupta, R.M. Kadam, N.S. Dhoble, S.P. Lochab and S.J. Dhoble, *Journal of Luminescence* 188 (2017) 81–95.
47. Structural and spectroscopic characterizations of a new near-UV converting cyan-emitting $RbBaScSi_3O_9:Eu^{2+}$ phosphor with robust thermal performance, Sudeshna Ray, Prachi Tadge, S.J. Dhoble, Govind B. Nair, Akash Singh, Abhishek Kumar Singh, Monika Rai, Teng Ming Chen, Vinita Rajput, *Journal of Alloys and Compounds*, 713 (2017) 138-147.
48. Preparation and tunable luminescence of microcrystalline $SrMg_2Al_{16}O_{27}:Tb^{3+}$and Eu^{3+} phosphor for solid state lighting, V.R. Panse, A. Shukla, S.V. Panse, N.S. Dhoble, S.J. Dhoble, *Journal of Materials Science: Materials in Electronics*, 28(2)(2017)2193-2199.
49. Urea assisted self combustion synthesis of $CaAl_2O_4$:Eu phosphor and its mechanoluminescence characterization, Girish C. Mishra, Kabita Ku. Satapathy, Sanjay J. Dhoble and Rajiv S. Kher, *New J. Chem.*, 41(2017)2193-2197.

50. Synthesis and luminescence characterization of $LaBO_3:Dy^{3+}$ phosphor for stress sensing application, Renu Nayar, Sumedha Tamboli, A.K. Sahu, Vinit Nayar and S.J. Dhoble, *J Fluoresc*, 27(2017) 251-261.
51. Investigation of thermoluminescence and electron-vibrational interaction parameters in $SrAl_2O_4:Eu^{2+}$, Dy^{3+} phosphors, Shilpa A. Pardhi, Govind B. Nair, Ravi Sharma and S.J. Dhoble, *Journal of Luminescence*, 187 (2017) 492–498.
52. Gadolinium-activated $CaZr_4O_9$ ultraviolet-B-emitting phosphor: A luminescence and EPR study,Vijay Singh, N. Singh, S. Watanabe, T.K. Gundurao,M.S. Pathak, A.K. Srivastava, P.K. Singh and S.J. Dhoble, *Journal of Electronic Materials*, 46(4)(2017)1943-1947.
53. Exploration of spectroscopic properties of solvated tris (thenoyltrifluoroacetonate) (2,2′-bipyridine)europium(III)red hybrid organic complex for solution processed OLEDs and displays, Dipti Chitnis, N. Thejokalyani, S.J. Dhoble, *Journal of Luminescence* 185 (2017) 61–71.
54. Portrayal of structural, thermal and optical properties of Ph sensitive Eu(TTA)3bipy hybrid organic complex for OLEDs, Dipti Chitnis, N. Thejo Kalyani, S.J. Dhoble, *Optik*, 130 (2017) 237–244.
55. Assessment of electron-vibrational interaction (EVI) parameters of $YAG:Ce^{3+}$, $TAG:Ce^{3+}$ and $LuAG:Ce^{3+}$ garnet phosphors by spectrum fitting method, Govind B. Nair, S.J. Dhoble, *Spectrochimica Acta Part A: Molecular and Biomolecular Spectroscopy*, 173 (2017) 822–826.
56. Effect of the synthesis route on the luminescence properties of Eu^{3+} activated $Li_6M(BO_3)_3$ (M = Y, Gd) phosphors, Mrunal M. Yawalkar, Govind B. Nair, G.D. Zade and S.J. Dhoble, *Materials Chemistry and Physics*, 189 (2017) 136-145.
57. Luminescence investigation of microcrystalline $Ca_2PO_4Cl:Dy^{3+}$ phosphor for eco-friendly solid state lighting, V.R. Panse, Alok shukla, S.J. Dhoble, *Optik*, 130 (2017) 539–542.

58. Luminescence of Cu^{+} and Mn^{3+} in $KMgCl_3$ nanomaterial, Anuradha Poddar, S.C. Gedam, S.J. Dhoble, *Journal of Luminescence*, 183 (2017) 73–78.
59. Photoluminescence properties of different luminescent ions (Ce^{3+}, Eu^{3+}, Eu^{2+}, Dy^{3+}, Cu^{+}) in K_2LaCl_5 host matrix, Priyanka D. Bhoyar, Govind B. Nair, S.J. Dhoble, *Optik*. 134 (2017) 33–44.
60. Synthesis and photoluminescence properties of Li_2SO_4: RE (RE = Eu^{3+}, Tb^{3+}, Gd^{3+} and Ce^{3+}) phosphors, D.R. Taikar, Sumedha Tamboli, S.J. Dhoble, *Optik* 139 (2017) 111–122.
61. On the infrared emissions of Er^{3+}/Yb^{3+} co-doped $SrAl_2O_4$,S.U. Pawar, Rangarajulu Senthil Kumarn, M. Kumar, K.S. Yun, S.J. Dhoble, Vijay Singh, *Optik* 136 (2017) 581–585.
62. Visible upconversion in Er^{3+}/Yb^{3+} co-doped $LaAlO_3$ phosphors, Vijay Singh, V.K. Rai, N. Singh, M.S. Pathak, M. Rathaiah, V. Venkatramu, Rahul V. Patel, Pramod K. Singh, S.J. Dhoble, *Spectrochimica Acta Part A: Molecular and Biomolecular Spectroscopy* 171 (2017) 229–235.
63. Optical and EPR spectroscopic studies of deep red light emitting Fe-doped $LiAl_5O_8$ phosphor prepared via propellant 5 combustion route, Vijay Singh, G. Sivaramaiah, J.L. Rao, N. SIingh, M. Mohapatra, P.K. Singh, M.S. Pathak and S.J. Dhoble, *Journal of Electronic Materials* 46(3), (2017), 1525–1531.
64. Efficient resonance energy transfer study from Ce^{3+} to Tb^{3+} in $BaMgF_4$, Bhushan P. Kore, Sumedha Tamboli, N.S. Dhoble, A. K Sinha, M.N Singh, S.J. Dhoble and H. C Swart, *Materials Chemistry and Physics* 187, (2017), 233-244.
65. Probing the thermodynamic and magnetic properties of UV-B emitting $GdAlO_3$ phosphors by ESR and optical techniques, Vijay Singh, G. Sivaramaiah, M. Mohapatra, J.L. Rao, N. Singh, M.S. Pathak, P.K. Singh, and S.J. Dhoble, *Journal of Electronic Materials* 46(2), (2017), 1137–1144.
66. Analysis of electron-vibrational interaction in the 5d states of Eu^{2+} ions in $A_3P_4O_{13}$ and $A_3(PO_4)_2$ (A= Sr, Ba) phosphors, Sumedha

Tamboli, P.D. Bhoyar and S.J. Dhoble, *Journal of Luminescence* 184, (2017), 23-28.

67. NUV excited $K_2SrP_2O_7:RE^{3+}$ (RE= Sm,Tb, Eu, Dy) phosphors for white light generation, Ketki S. Dhoble, J.A. Wani and S.J. Dhoble, *Advanced Materials Letters*, 7(9)(2016) 765-769.
68. Comparison of thermoluminescence characteristics in γ-ray and C^{5+} ion beam-irradiated $LiCaAlF_6$:Ce phosphor, M.M. Yerpude, N.S. Dhoble, S.P. Lochab and S.J. Dhoble, *Luminescence: The Journal of Biological and Chemical Luminescence*, 31(2016) 1115-1124.
69. Escalating opportunities in the field of lighting, Dipti Chitnis, N. Thejo kalyani, H.C. Swart and S.J. Dhoble, *Renewable & Sustainable Energy Reviews* 64(2016) 727-748.
70. Photoluminescence, thermoluminescence and evaluation of some parameters of Dy^{3+} activated $Sr_5(PO_4)_3F$ phosphor synthesized by sol-gel method, Karan Kumar Gupta, R.M. Kadam, N.S. Dhoble, S.P. Lochab, Vijay Singh and S.J. Dhoble, *Journal of Alloys and Compounds*, 688(2016)982-993.
71. Synthesis and characterization of novel $Na_{15}(SO_4)_5F_4Cl:Ce^{3+}$ halosulfate phosphors, A.M. Bhake, Govind B. Nair, G.D. Zade and S.J. Dhoble, *Luminescence: The Journal of Biological and Chemical Luminescence*, 31(2016)1468-1473.
72. Cr^{3+}-doped $Yb_3Ga_5O_{12}$ nanophosphor: Synthesis,Optical, EPR, Studies,Vijay Singh, G. Sivaramaiah, J.L. Rao, N. Singh, M.S. Pathak, H.D. Jirimali, Pramod K. Singh,Anoop K. Srivastava, S.J. Dhoble and M. Mohapatra, *Journal of Electronic Materials*, 45(8)(2016)4076-4082.
73. Combustion synthesized Cr^{3+}-doped-$BaMgAl_{10}O_{17}$ phosphor: An electron paramagnetic resonance and optical study, Vijay Singh, G. Shivramaiah, J.L. Rao, Anoop K. Srivastava, R.V.S.S.N. Ravikumar, S.J. Dhoble, P.K. Singh and Manoj Mohapatra, *J. Electronic Materials*, 45(1) (2016) 365-373.
74. Stabilization of ZnS nanoparticles by polymeric matrices: Syntheses, optical properties and recent applications, Ashish Tiwari and S.J. Dhoble, *RSC Advances*, 6(2016)64400–64420.

75. EPR and optical properties of Eu^{2+} and Mn^{2+} co-doped $MgSrAl_{10}O_{17}$ blue-green light emitting powder phosphors, N. Singh, Vijay Singh, G. Sivaramaiah, J.L. Rao, Pramod K. Singh, M.S. Pathak, S.J. Dhoble and M. Mohapatra, *J Lumin* 178 (2016) 479-486.
76. Photoluminescence and ESR identification of γ-ray irradiation induced defects responsible for thermoluminescence in Tb^{3+} activated hibonite ($CaAl_{12}O_{19}$) powder material, Vijay Singh, S. Watanabe, T.K. Gundu Rao, N. Singh, Anoop K. Srivastava, Pramod K. Singh, H. Gao, P. Mardina and S.J. Dhoble, *Journal of Electroceramics*, 37(1-4)(2016)58-65.
77. Studies on the luminescence properties of the gamma-ray-irradiated white light emitting $Ca_2Al_2SiO_7$:Dy^{3+} phosphors by combustion-assisted method, Geetanjali Tiwari, Nameeta Brahme, Ravi Sharma, D.P. Bisen, Sanjay K. Sao and S.J. Dhoble, *RSC Advances*, 6(2016)49317–49327.
78. Luminescence studies on $BaSi_6N_8O$: Eu^{3+} phosphors, S.A. Fartode, B. Sudhakar Reddy, V. Nayar and S.J. Dhoble, *Ferroelectric Letters*, 43(2016)34-51.
79. Crystal structure, energy transfer mechanism and tunable luminescence in Ce^{3+}/Dy^{3+} co-activated $Ca_{20}Mg_3Al_{26}Si_3O_{68}$ nanophosphors, Ashwini Kumar, S.J. Dhoble, D.R. Peshwe, Jatin Bhatt, J.J. Terblans and H.C. Swart, *Ceramics International*, 42(2016) 10854-10865.
80. Photoluminescence investigation of trivalent rare earth activated $Na_3Pb_2(SO_4)_3Cl$ phosphors for solid state lighting, Nahida Baig, N.S. Dhoble, A.N. Yerpude, Vijay Singh, and S.J. Dhoble, *Optik*, 127(2016)6574-6582.
81. Quenching and dipole–dipole interactions in $Sr_2Al_2SiO_7$: Ce^{3+} host lattice, M. Kolte, V.B. Pawade and S.J. Dhoble, *Applied Physics A: Materials Science and Processing*, 122(2016)59.
82. Combustion synthesis of rare earth activated and co-activated $SrAl_4O_7$ green long lasting phosphors, A.N. Yerpude, S.J. Dhoble, *Optik* 127 (2016) 4217–4221.

83. Luminescence study of Eu^{3+} doped $Li_6Y(BO_3)_3$ phosphor for solid state lighting, Mrunal M. Yawalkar, G.D. Zade, K.V. Dabre and S.J. Dhoble, *Luminescence: The Journal of Biological and Chemical Luminescence*, 31(4)(2016)1037-1042.

84. Energy transfer between $Ce^{3+} \rightarrow Gd^{3+}$ or Tb^{3+} in $KNaSO_4$ microphosphor, Urvashi Manik, S.C. Gedam and S.J. Dhoble, *Luminescence: The Journal of Biological and Chemical Luminescence*, 33(3)(2016) 911-914.

85. Effect of Li^+ and Zn^{2+} co-doping on PL properties of Y_2O_3:Eu^{3+} phosphor, R.S. Ukare, G.D. Zade, B. Deva Prasad Raju and S.J. Dhoble, *Optik* 127 (2016) 1871–1878.

86. Blue and green emitting Eu^{2+}, Ce^{3+} and Tb^{3+} activated novel $BaAl_2SiO_6$ phosphor, M. Kolte, V.B. Pawade, S.J. Dhoble, *Optik* 127 (2016) 1703–1707.

87. Combustion synthesis of $Sr_2B_2O_5$:Tb^{3+} green emitting phosphor for solid state lighting, V.R. Panse, N.S. Kokode, A.N. Yerpude and S.J. Dhoble, *Optik* 127 (2016) 1603–1606.

88. Combustion synthesized europium doped $LaAl_{11}O_{18}$ phosphors—An electron paramagnetic resonance and optical study, Vijay Singh, G. Sivaramaiah, J.L. Rao, Kumaran, R. Senthil and S.J. Dhoble, *Journal of Nanoscience and Nanotechnology*, 16(1)(2016) 789-796.

89. Tunable luminescence properties and energy transfer in $LaAl_{11}O_{18}$:Eu,Tb phosphor, M.S. Mendhe, S.P. Puppalwar and S.J. Dhoble, *Luminescence: The Journal of Biological and Chemical Luminescence*, 33(3)(2016)881-887.

90. Eu^{2+} and Mn^{2+} co-doped $BaMgAl_{10}O_{17}$ blue- and green-emitting phosphor: A luminescence and EPR study, Vijay Singh, G. Sivaramaiah, J.L. Rao, N. Singh, Anoop K. Srivastava, H.D. Jirimali, J. Li, H. Gao, R. Senthil Kumaran, P.K. Singh and S.J. Dhoble, *Journal of Electronic Materials*, 45(6) (2016) 2776-2783.

91. PL properties of Sr_2CeO_4 with Eu^{3+} and Dy^{3+} for solid state lighting prepared by precipitation method, R.S. Ukare, Vikas Dubey, G.D. Zade and S.J. Dhoble, *Journal of Fluorescence*, 26(2016) 791–806.

92. Effect of Ce^{3+} ion on Dy^{3+} or Mn^{2+} in $KMgSO_4Cl$ synthesized by centrifuge method, Nita Shinde, N.S. Dhoble, S.C. Gedam, S.J. Dhoble, *Journal of Luminescence*,172(2016)131–138.
93. Thermoluminescence and EPR study of $K_2CaMg(SO_4)_3$:Dy phosphor: the dosimetric application point of view, Bhushan P Kore, N S Dhoble, R M Kadam, S P Lochab, M N Singh, S J Dhoble and H C Swart, J. *Phys. D: Appl. Phys*. 49 (2016) 095102.
94. EPR and optical properties of green emitting Mn-doped $BaMgAl_{10}O_{17}$ nano-phosphors prepared by a combustion reaction, Vijay Singh, M. Mohapatra, G. Sivaramaiah, J.L. Rao, N. Singh, H. Gao, J. Li, Anoop K. Srivastava, H.D. Jirimali, S.J. Dhoble, Pramod K. Singh, K.V.R. Murthy and V. Natarajan, *J Mater Sci: Mater Electron* (2016) 27:3697–3703.
95. Synthesis and characterization of $Ba_3Si_6O_{12}N_2:Eu^{2+}$ phosphor, S.A. Fartode and S.J. Dhoble, *Luminescence: The Journal of Biological and Chemical Luminescence*, 31(1)(2016) 295–304.
96. Synthesis and characterization of erbium-doped $YAlO_3$ phosphor, Huma Nazli Baig, Jagjeet Kaur Saluja and S.J. Dhoble, *Luminescence: The Journal of Biological and Chemical Luminescence*, 31(2), (2016) 401-406.
97. A novel red-emitting phosphor, $KAl_{1-x}PO_4Cl:Eu_x^{3+}$ ($0.1 \leq x \leq 1.0$), Ashish Akojwar, K.N. Shinde, Roshani Singh, S.J. Dhoble and N.S. Kokode, *Luminescence: The Journal of Biological and Chemical Luminescence*, 31(1) (2016) 305–308.
98. Synthesis, characterization and optical properties of polymer-based ZnS nanocomposites, A. Tiwari, S.A. Khan, R.S. Kher, S.J. Dhoble and A.L.S. Chandel, *Luminescence: The Journal of Biological and Chemical Luminescence*, 31(2) (2016) 428-432.
99. Investigation of UV-emitting Gd^{3+}-doped $LiCaBO_3$ phosphor, Sumedha Tamboli, B. Rajeswari and S.J. Dhoble, *Luminescence: The Journal of Biological and Chemical Luminescence*, 31(2)(2016) 551–556.
100. Synthesis and photoluminescence characteristics of $(Y,Gd)BO_3$:RE (RE=Eu^{3+}, Ce^{3+}, Dy^{3+} and Tb^{3+}) phosphors for blue chip and near-

135. Luminescence of Eu and Ce in $K_3Ca_2(SO_4)_3F$ fluoride material, Anuradha Poddar, S.C. Gedam and S.J. Dhoble, *Luminescence: The Journal of Biological and Chemical Luminescence*, 30(6)(2015)914-917.
136. Synthesis and thermoluminescence characterizations of $Sr_2B_5O_9Cl:Dy^{3+}$ phosphor for TL dosimetry, Abha H. Oza, N.S. Dhoble, K. Park and S.J. Dhoble, *Luminescence: The Journal of Biological and Chemical Luminescence*, 30(6)(2015)768-774.
137. Characterization, luminescence and defect centres in Ce^{3+} doped $Li_2Si_2O_5$ phosphor obtained by solution combustion reaction, Vijay Singh, S. Watanabe, T.K. Gundu Rao, S.J. Dhoble, *Journal of Electronic Materials*, 44 (8), (2015), 2736-2744.
138. Photoluminescence enhancement in Na_3SO_4Cl: X (X= Ce^{3+}, Eu^{3+} or Dy^{3+}) material, Nita Shinde, N.S. Dhoble, S.C. Gedam and S.J. Dhoble, *Luminescence: The Journal of Biological and Chemical Luminescence* 30(6)(2015)893-903.
139. Electron Paramagnetic Resonance and Photoluminescence Studies of $LaMgAl_{11}O_{19}$: Mn^{2+} Green Phosphors, V Singh, R P S Chakradhar, J L Rao, S J Dhoble, S H Kim, *Journal of Electronic Materials* 43 (11)(2015) 4041-4047.
140. A review on optical and photoluminescence studies of RE^{3+} (RE = Sm, Dy, Eu, Tb and Nd) ions doped LCZSFB glasses, B. Deva Prasad Raju, C. Madhukar Reddy, N. John Sushma, N.S. Dhoble and S.J. Dhoble, *Renewable & Sustainable Energy Reviews*, 51(2015)566–584.
141. Photoluminescence properties of $LiBaPO_4:M^{3+}$ phosphor for near-UV light-emitting diode (M=Eu and Dy), S.P. Puppalwar and S.J. Dhoble, *Luminescence: The Journal of Biological and Chemical Luminescence*, 30(6)(2015)745-750.
142. Synthesis and assessment of photoluminescent properties of $Ca_{4-2x}Al_6WO_{16}:RE_x,Na_x$ (RE $=Eu^{3+}$,Dy^{3+} and Sm^{3+}) phosphors, K.V. Dabre and S.J. Dhoble, *RSC Adv.*, 5(2015)60409–60418.
143. Electron paramagnetic resonance and photoluminescence investigation on ultraviolet-emitting gadolinium-ion-doped $CaAl_{12}$

O_{19} phosphors, Vijay Singh, S. Borkotoky, A. Murali, J.L. Rao, T.K. Gundu Rao and S.J. Dhoble, *Spectrochimica Acta Part A: Molecular and Biomolecular Spectroscopy* 139 (2015) 1–6.

144. A high sensitive phosphor for dosimetric applications, Bhushan P. Kore, N.S. Dhoble, S.P. Lochab and S.J. Dhoble, *AIP Conference Proceedings* 1665(2015)140029-1- 140029-3.
145. Highly enterprising calcium zirconium phosphate [$CaZr_4(PO_4)_6$: Dy^{3+}, Ce^{3+}] phosphor for white light emission, Govind B. Nair and S.J. Dhoble, *RSC Advances*, 5,(2015)49235–49247.
146. A comparative investigation of γ-ray and C^{5+} ion beam impact on thermoluminescence response of Mg_2BO_3F:Dy phosphor, Bhushan P. Kore, N.S. Dhoble, R.M. Kadam, S.P. Lochab and S.J. Dhoble, *Materials Chemistry and Physics* 161 (2015) 96-106.
147. Synthesis and optical characterization of $Ca_2PO_4Cl:Tb^{3+}$ and Mn^{3+} phosphor for solid state lighting, N.S. Kokode, V.R. Panse and S.J. Dhoble, *Advanced Materials Letters*,6(2015)616-619.
148. Synthesis of $Sr_5SiO_4Cl_6:Eu^{3+}$, $NaAlSiO_4:Tb^{3+}$ and $Sr_2P_2O_7$: Dy^{3+} phosphors for solid state lighting applications, S.J. Dhoble, Bhushan P. Kore, A.N. Yerpude, R.L. Kohale, P.W. Yawalkar and N.S. Dhoble, *Optik* - International Journal for Light and Electron Optics, 126 (2015) 1527–1533.
149. Synthesis, photoluminescence and Judd–Ofelt parameters of $LiNa_3P_2O_7:Eu^{3+}$ orthorhombic microstructures, K. Munirathnam, G.R. Dillip, B. Deva Prasad Raju, S.W. Joo, S.J. Dhoble, B.M. Nagabhushana, R. Hari Krishna, K.P. Ramesh, S. Varadharaj Perumal, D. Prakashbabu, *Applied Physics A*,120,(2015)1615-1623.
150. Thermoluminescence of γ- irradiated $BaCa(SO_4)_2$: Eu, Dy, V.C. Kongre, S.C. Gedam, S.J. Dhoble, *Radiation Effects and Defects in Solids*, 170(7-8)(2015)610-620.
151. Synthesis and thermoluminescence characteristics of γ - irradiated $K_3Ca_2(SO_4)_3F$: Eu or Ce Fluoride, Anuradha Poddar, S.C. Gedam and S.J. Dhoble, *Radiation Effects and Defects in Solids*, 170(5)(2015)451-460.

152. Photoluminescence characteristics of Ce and Eu activated $MBPO_5$ (M= Sr, Ca) phosphors, V.R. Kharabe, Abha H. Oza and S.J. Dhoble, *Optik*, 126 (2015) 4544-4547.
153. Mechanoluminescence and thermoluminescence characterization of Tb^{3+} doped $CaAl_2O_4$: a theoretical and experimental study, K.K. Satapathy, G.C. Mishra, R.S. Kher and S.J. Dhoble, *RSC Adv.*, 5 (2015) 79391-79396.
154. Synthesis and characterization of trivalent RE (RE = Eu^{3+}, Dy^{3+}, Ce^{3+}) doped new $Ca_3Al_2Si_3O_{12}$ materials for NUV-wLEDs, V.B. Pawade, N.S. Dhoble and S.J. Dhoble, *Mater. Res. Express*, 2 (2015) 095501.
155. Optical property investigations of polystyrene capped $Ca_2P_2O_7$: Dy^{3+} persistent phosphor, Abhilasha Jain, Ashwini Kumar, S.J. Dhoble and D.R. Peshwe, *Materials Research Bulletin*, 70(2015)980-987.
156. Electron paramagnetic resonance and photoluminescence studies of ultraviolet emitting $ZnAl_2O_4$:Gd^{3+} phosphors, Vijay Singh, G. Sivaramaiah, J.L. Rao, S.J. Dhoble and S.H. Kim, *Journal of Electronic Materials*, 44(1)(2015)121-128.
157. Mn^{2+}, Eu^{2+} and Eu^{3+} emission in co-doped $LaAl_{11}O_{18}$ phosphors, Vijay Singh, G. Sivaramaiah, J.L. Rao, S.J. Dhoble and S.H. Kim, *Materials Chemistry and Physics*, 149-150 (2015) 202-208.
158. Effect of γ– radiation on thermoluminescence in rare earths doped $NaMgSO_4Cl$ material, S.R. Choubey, S.C. Gedam and S.J. Dhoble, *Radiation Effects and Defects in Solids*, 170(1)(2015)7-17.
159. Thermoluminescence Investigations of sol-gel derived and γ-irradiated rare earth (Eu and Nd) doped YAG nanophosphors, M.S. Kurrey, Ashish Tiwari, M.S.K. Khokhar, R.S. Kher and S.J. Dhoble, *Journal of Luminescence*, 164 (2015) 94–98.
160. Combustion synthesis and luminescence characterization of Ce^{3+} and Dy^{3+} activated $K_2Ca(WO_4)_2$ novel phosphors, Parag Nimishe, K. Park and S.J. Dhoble, *Combustion Science and Technology*, 187(2015)953–962.
161. Luminescence investigations of RE^{3+} (RE^{3+}=Dy^{3+}, Eu^{3+}, Tb^{3+}) activated $Na_{21}(SO_4)_7F_6Cl$ optoelectronics nanophosphors under near

UV excitation for LED,Vilas V. Shinde, Shalaka V. Shinde, N.S. Dhoble and S.J. Dhoble, *Journal of Inorganic and Organometallic Polymers and Materials*, 25(2015) 593–600.

162. Photo and thermoluminescence in $KMgSO_4F$:Ce or Mn phosphor, A. Poddar, S.C. Gedam and S.J. Dhoble, *Luminescence: The Journal of Biological and Chemical Luminescence*, 30(2015) 425–431.
163. Luminescence in $Sr_4Al_{14}O_{25}:Ce^{3+}$ aluminate phosphor, G.N. Nikhare, S.C. Gedam, and S.J. Dhoble, *Luminescence: The Journal of Biological and Chemical Luminescence*, 30(2015) 163–167.
164. Luminescent properties of $MAl(SO_4)_2Br:Eu^{3+}$ (M=Sr or Mg) red phosphors for near-UV light-emitting diodes, Priti B. Deshmukh, S.P. Puppalwar, N.S. Dhoble, S.J. Dhoble, *Luminescence: The Journal of Biological and Chemical Luminescence*, 30(2015)118–121.
165. Stability of luminescence in $LaPO_4$, $LaPO_4:RE^{3+}$ (RE= Dy, Eu) nanophosphors, D.M. Pimpalshende and S.J. Dhoble, *Luminescence: The Journal of Biological and Chemical Luminescence*, 30(2015)144–154.
166. Synthesis, PL characterizations and concentration quenching effect in Dy^{3+} and Eu^{3+} activated $LiCaBO_3$ phosphors, V.R. Kharabe, Abha H. Oza and S.J. Dhoble, *Luminescence: The Journal of Biological and Chemical Luminescence*, 30(2015)432–438.
167. Novel Br-DPQ blue light emitting phosphors for OLED, H.K. Dahule, N. Thejo Kalyani, S.J. Dhoble, *Luminescence: The Journal of Biological and Chemical Luminescence*, 30(2015)405–410.
168. Luminescence characteristics of C^{5+} ions and ^{60}Co irradiated $Li_2BaP_2O_7$: Dy^{3+} phosphor, J.A. Wani, N.S. Dhoble, S.P. Lochab and S.J. Dhoble, *Nuclear Inst. and Methods in Physics Research, B*, 349 (2015) 56–63.
169. Luminescence study of γ- ray and C^{5+} ion beam irradiated $LiCaBO_3$: Cu phosphor, Abha H. Oza, N.S. Dhoble, S.P. Lochab and S.J. Dhoble, *Radiation Effects and Defects in Solids*, 170(7-8)(2015)659-667.

170. Review of rare earth activated blue emission phosphors prepared by combustion Synthesis, V.B. Pawade, H.C. Swart and S.J. Dhoble, *Renewable & Sustainable Energy Reviews*, 52(2015)596-612.
171. Enhanced luminescence and white light emission from Eu^{3+}-co-doped $K_3Ca_2(SO_4)_3Cl:Dy^{3+}$ phosphor with near visible ultraviolet excitation for white LEDs, N. Baig, N.S. Dhoble, K. Park, N.S. Kokode and S.J. Dhoble, *Luminescence: The Journal of Biological and Chemical Luminescence*, 30(2015)479–484.
172. Novel Na^{+} doped Alq_3 hybrid materials for organic light-emitting diode (OLED) devices and flat panel displays, S.A. Bhagat, S.V. Borghate, N. Thejo Kalyani and S.J. Dhoble, S.J. Dhoble, *Luminescence: The Journal of Biological and Chemical Luminescence,* 30(2015) 251–256.
173. Luminescence characteristics of $Na_3Ca_2(SO_4)_3F$: Ce^{3+} fluoride material,G.N. Nikhare, S.C. Gedam, S.J. Dhoble, *Luminescence: The Journal of Biological and Chemical Luminescence*, 30(5)(2015)703-706.
174. Photoluminescence of Dy^{3+} activator sensitized by Ce^{3+} in $KNaSO_4$ microphosphor, Urvashi Manik, S.C. Gedam and S.J. Dhoble, Luminescence: *The Journal of Biological and Chemical Luminescence*, 30(6)(2015)910-913.
175. Study of luminescence and effect of Dy on $NaMgSO_4Cl$: Ce chlorosulphate phosphor, S.R. Choubey, S.C. Gedam and S.J. Dhoble, Luminescence: *The Journal of Biological and Chemical Luminescence*, 29(8)(2015)1199-1201.
176. Emission analysis of RE^{3+}(RE=Eu,Sm,Dy):$MgY_4Si_3O_{13}$ phosphors, B. Sudhakar Reddy, Monali R. Kadukar, Hyeong-Yong Hwang, Byoung Seung Ham, S. Sailaja, U. Rambabu, S.J. Dhoble, Young-Dahl Jho, *Physica B*, 463 (2015) 39–47.
177. Influence of P ion on $Sr_2B_5O_9Cl$:Eu for TL dosimetry, Abha H. Oza, N.S. Dhoble and S.J. Dhoble, *Nuclear Inst. and Methods in Physics Research, B*. 344 (2015) 96–103.
178. Optical properties of Nd^{3+}-doped and Er^{3+}- Yb^{3+} codoped borotellurite glass for use in NIR lasers and fiber amplifiers, B.

Sudhakar Reddy, Hyeong-Yong Hwang,Young-Dahl Jho,Byoung Seung Ham, S. Sailaja, C. Madhukar Reddy, B. Vengala Rao and S.J. Dhoble, *Ceramics International*, 41 (2015) 3684–3692.

179. Combustion synthesis of $SrAl_4O_7:Eu^{2+}$, Dy^{3+}, Gd^{3+} long lasting phosphor, A.N. Yerpude and S.J. Dhoble, *IOP Conf. Series: Materials Science and Engineering*, 73 (2015) 012126.
180. Blue-green luminescence in Hg free excited $Sr_2P_2O_7:Tb^{3+}$ pyrophosphate phosphor for NUV excited LEDs, R.L. Kohale, O.P. Chimankar and S.J. Dhoble, *IOP Conf. Series: Materials Science and Engineering*, 73 (2015) 012111.
181. Eu and Dy activated $(Sr_{0.90}Ca_{0.10})_3(PO_4)_2$ phosphors for near- UV LED, K.N. Shinde and S.J. Dhoble, *IOP Conf. Series: Materials Science and Engineering*, 73 (2015) 012103.
182. Wet chemical synthesis of $Na_{21}(SO_4)_7F_6Cl:Ce$ optoelectronics materials, V.V. Shinde and S.J. Dhoble, *IOP Conf. Series: Materials Science and Engineering*, 73 (2015) 012038.
183. Synthesis and photoluminescence studies of Ba(Gd,Ln) $B_9O_{16}:Eu^{3+}$(Ln= La,Y) phosphors for n-UV LED lighting and display devices, V.V. Rangari and S.J. Dhoble, *Journal of Rare Earth*, 33(2)(2015) 140- 146.
184. Novel materials for fabrication and encapsulation of OLEDs, N. Thejo Kalyani and S.J. Dhoble, *Renewable & Sustainable Energy Reviews*, 44(2015)319–347.
185. Synthesis of $KMgCl_3$ nanomaterial and luminescence of $Ce^{3+}/Dy^{3+}/Eu^{3+}$ by different routes, Anuradha Poddar, S.C. Gedam and S.J. Dhoble, *Journal of Luminescence*,158 (2015) 188–196.

In: Ionizing Radiation
Editor: Tamar Reeve
ISBN: 978-1-53613-539-8

Chapter 2

SYNERGISTIC APPLICATIONS OF IONIZING RADIATION, RADIATION-INDUCED CHANGES IN PRE-PACKAGED FOODS AND ELECTRONIC SENSING-BASED DISCRIMINATION OF IRRADIATED FOODS

Yunhee Jo, Kashif Ameer, Sudha Rani Ramakrishnan and Joong-Ho Kwon
School of Food Science and Biotechnology,
Kyungpook National University, Daegu, South Korea

ABSTRACT

Research has established food irradiation as a technically feasible option in the food industries. Irradiation is a cold process that improves the safety and quality of food commodities. Ionizing radiations, such as gamma ray, electron beam and X-rays have been approved for decontaminating packaged/frozen foods, feeds, and packaging materials. These methods are particularly effective in destroying harmful organisms in raw and processed materials. Approval of irradiation treatments in the food sector is linked with regulation-compliant labelling, availability of detection methods, and international trade regulations. Recent advances regarding effects on exposure of some foods to ionizing radiations are presented. The dosage required for complete sterilization may, at times, lead to undesirable changes in food flavours or may exceed the permitted levels. Combining irradiation with other treatments yields satisfactory results in these cases. Combined applications of ionizing radiation with heat, low temperature, high hydrostatic pressure, and modified atmospheres are also discussed. Recently, safety of packaging systems has also been taken into consideration to guarantee the safety of using radiation for sterilization of food and healthcare products. Irradiation of pre-packaged foodstuffs may generate radiolytic by-products in the packaging material, which may compromise their barrier properties and allow the radiolytic products to enter the foodstuffs. Technical challenges regarding the safety of food packaging materials have also been elaborated. Moreover, innovative electronic sensing techniques, such as electronic nose and electronic tongue, for authentication and control of irradiated foods are discussed for potential future applications.

Introduction

Globalization of the food market has resulted in the tremendous growth of international trade. Quarantine requirements and quality of merchandized commodities are the most important issues among trading countries. Many methods have been traditionally used for improving food quality and safety, including heat or cold treatments, chemicals, and atmospheric modifications to control microbes, pests, and physiological changes. New alternative technologies, however, are needed to overcome the limitations of existing methods, including technical limits and hazards to environment and human health posed by the use of chemicals (WHO, 1999).

Food irradiation, a cold pasteurization technique, is gaining attention in the global food sector as one of the alternative methods, owing to its safety and technical efficacy, and has been investigated thoroughly for more than half a century (Roberts, 1998). Furthermore, food irradiation is increasingly being deemed as a solution for persistent problems such as

>25% of postharvest food losses and foodborne diseases. Moreover, increasing restrictions on the use of various chemical fumigants make food irradiation a lucrative alternative for protecting foodstuffs against pests and harmful microorganisms (Table 1).

Thermal processing for assurance of food safety has been a widely-adopted method in the food industry. However, until recently, thermal processing was associated with adverse effects, such as undesirable changes or formation of by-products that negatively affected the nutritional value, flavour, colour, and texture of the final product. Health conscious consumers are inclined towards naturally and minimally processed foods. Therefore, exploring non-thermal processing technologies that have minimal impact on food but also efficiently extend the product shelf-life by reducing microbial contamination was inevitable. Thus, irradiation, a type of non-thermal processing, gained popularity in recent times.

Non-thermal microbial inactivation technologies are based on subjecting foodstuffs to electromagnetic fields, pulsed electric fields (PEF), high voltage arc discharge (HVAD), pulsed light, ionizing radiation, microwave, and cold plasma. Plasma is a partially-ionized mixture of reactive species, such as electrons, positive and negative ions, free radicals, gas atoms, and photons (Pankaj et al. 2014), which inactivates microbial population in food (Mai-Prochnow et al. 2014).

Further, microwave, electric or electromagnetic fields, and X-rays are useful tools for gas ionization (Stoica et al. 2014). Microwaves can be used to ionize a gas into plasma without the need of electrodes (Bárdos and Baránková, 2010). The application of HVAD could significantly increase the aqueous and hydro-ethanolic yields of the total phenolic content (Roselló-Soto et al. 2015), while PEF and irradiation showed minimal influence on the anthocyanin content of fruit juices (Tiwari et al. 2009). Irradiation at 1.0 kGy reduced the level of pesticides in foods; prominent examples include reduction of chlorpyrifos in cucumber, diazinon in capsicum, and phosphamidon in tomato by 80-91%, 85-90%, and 90-95%, respectively (Chowdhury et al. 2014). A common feature of these technologies is that no active heat is applied to the food products, implying that the temperature is low or moderate (below 60°C).

Table 1. Recent studies on food irradiation based on foods permitted to be irradiated with a technical dose limit stipulated in the WHO/FAO/IAEA & Codex Alimentarius regulations

Classes of food	Purpose	Maximum dose (kGy)	References
Class 1: Bulbs, roots and tubers	To inhibit sprouting during storage	0.2	Abboudi et al. (2016) Ichinose et al. (2016) Lee et al. (2016) Blessington et al. (2015) Mahto and Das (2015) Sadoughi et al. (2015) Janagam et al. (2014) Mahto and Das (2014) Sadoughi et al. (2013)
Class 2: Fresh fruits and vegetables (other than Class 1)	To delay ripening Insect disinfestation Shelf life extension Quarantine control	1.0 1.0 2.5 1.0	Jo et al. (2018) Abad et al. (2017) Finten et al. (2017) Hallman (2017) Tao et al. (2017) Cetinkaya et al. (2016) Cheon et al. (2016a) Jo et al. (2016) Pinela et al. (2016) Shahbaz et al. (2016) Jo et al. (2015) Rashid et al. (2015) Jouki and Khazaei (2014) Zhan et al. (2014)
Class 3: Cereals and their milled products, nuts, oil seeds, pulses, dried fruits	Insect disinfestation Reduction of microbial load	1.0 5.0	Aylangan et al. (2017) Luo et al. (2017) Bhat et al. (2016) Marathe et al. (2016) Khalil and Al-Bachir (2015) Severino et al. (2015) Severino et al. (2014b) Supriya et al. (2014)

Classes of food	Purpose	Maximum dose (kGy)	References
Class 4: Fish, seafood and their products (fresh or frozen)	Reduction of certain pathogenic micro-organisms Shelf life extension Control of infection by parasites	5.0 3.0 2.0	Ihsanullah and Rashid (2017) Kang et al. (2016) Arshad et al. (2015) Kim et al. (2015) Shaviklo (2015) Ahn et al. (2014) Hong and Kim (2014) Ahn et al. (2013) El-Nesr et al. (2013) Park et al. (2013)
Class 5: Raw poultry and meat and their products (fresh and frozen)	Reduction of pathogenic micro-organisms Shelf life extension Control of infection by parasites	7.0 3.0 2.0	An et al. (2017) Feng et al. (2016) Alahakoon et al. (2015) Huq et al. (2015) Li et al. (2015) Lung et al. (2015) Yong et al. (2015) Jung et al. (2014)
Class 6: Dry vegetables, spices, condiments, animal feed, dry herbs and herbal teas	Reduction of certain pathogenic micro-organisms Insect disinfestation	10.0 1.0	Fernandes et al. (2017) Kang et al. (2017) Khan et al. (2017) Khatun et al. (2017) Pillai and Shayanfar (2017) Vardhan and Shukla (2017) Fernandes et al. (2016) Fanaro et al. (2015) Kancheva et al. (2015) Marcazzó et al. (2015) Zaman et al. (2015) Fanaro et al. (2014) Huang et al. (2014) Kirkin et al. (2014) Tsai et al. (2014) Zhang et al. (2014) Kim et al. (2013) Polovka and Suhaj (2013)
Class 7: Dried food of animal origin	Disinfestation Control of moulds Sterilization Quarantine control	1.0 3.0 >10 >10	Zhao et al. (2018) Jeong et al. (2017) Feng and Ahn (2016) Domijan et al. (2015) Huq et al. (2015)

Potatoes, onions, garlic, chestnuts, mushrooms, yeast and algal foods, egg powder, grains, pulses, starches, dried meat products, fish powder, shellfish powder, crustacean powder, soybean paste, hot pepper paste, soy sauce (powdered), dried vegetables, ginseng, dried spices, and teas have been irradiated for various objectives. These include pest control, sprout inhibition, ripening control, and pasteurization. Details regarding the purpose of irradiation of these foodstuffs and the corresponding dosage used are given in Table 1.

This chapter introduces ionizing radiations used in food processing, and applications of food irradiation in combination with other treatments. In addition, radiation-induced changes in pre-packaged foods, effects on the barrier properties of the packaging material, technical challenges, and safety issues are also discussed in detail. Applications of electronic sensing techniques (electronic nose and electronic tongue) for screening of irradiated foods are also described in detail, along with safety issues of irradiated foods.

Effects of Irradiation Combined with Near-Infrared, Ultra-Violet, Mild Temperature, and Thermal Treatments

It is important to consider the presence of sublethally-injured cells after bactericidal treatments, as these cells can recover and proliferate under favourable conditions. Synergistic bacterial inactivation by simultaneous near-infrared (NIR) heating and UV treatment damages cellular envelopes, resulting in ribosomal damage, which the cells are unable to repair. Thus, more than one type of damage may be incurred with the combined use of lethal methods (Ha and Kang, 2015). The combination of UV light and mild temperatures could achieve 99.99% (4D) reduction of *Salmonella enterica* subsp. *enterica* (Gayán et al. 2012). The combination of irradiation with heat treatment has been shown to be an effective method to sensitize spores of *Clostridium thermosaccharolyticum* ATCC 7956 and *Desulfotomaculum nigrificans* ATCC 7946 (Pillai and Shayanfar, 2014). Marination combined with a low dose of γ-irradiation (1.5 kGy) reduced the pathogenic bacterial count to undetectable levels and increased the shelf-life of ready-to-cook meat loin without affecting its sensory and nutritional qualities (Fadhel et al. 2016).

Combination of Irradiation with High-Pressure Processing

Safe and convenient mutton patties without loss of natural flavour, freshness, appearance, and nutrient content can be obtained by using various combinations of processing techniques, such as high-pressure processing (200, 400 MPa), gamma irradiation (3 kGy), and vacuum packaging. In mutton patties treated with a combination of irradiation and pressure, the total microbial plate counts were reduced by 2–3 log CFU (Banerjee et al. 2017). The processed food experiences isostatic transmission of pressure instantaneously, without gradient, resulting in uniform treatment, irrespective of the size and geometry of the product. The treatment modifies only noncovalent bonds and does not affect small molecules, thereby preserving the flavour and vitamins. Electron-beam irradiation (1, 2 kGy) and high-pressure processing (300, 400 Mpa) maintained the quality characteristics of marinated chicken breast meat with and without citrus peel extract (Alahakoon et al. 2015). These minimal processing approaches retain their physicochemical, colour, textural, and sensory properties as well as reduce microbial counts of the meat products.

Irradiation Coupled with Antimicrobial Formulations

Microorganisms that survived after gamma irradiation combined with other non-thermal treatments have been shown to be more sensitive than untreated cells to adverse environmental conditions, such as the presence of antimicrobial compounds or modified atmospheric conditions (Severino et al. 2014a). Irradiation treatment after treating with natural antimicrobial formulations resulted in a short-term antimicrobial effect (Turgis et al. 2012; Takala et al. 2011; Yun et al. 2011). Therefore, spraying of natural antimicrobial formulations on products, followed by the irradiation (UV-C or gamma) treatment on products, prevents interferences between these two methods. The combination of gamma irradiation at 1 kGy with natural antimicrobials (oregano or lemongrass essential oil plus citrus extract and lactic acid) reduced the population of *L. monocytogenes* and *E. coli* O157:H7 below the detection limits during the storage period (Tawema et al. 2016). A mixture of carvacrol and nisin increased the radiosensitization of *L. monocytogenes* and *S. typhimurium* in mini carrots (Ndoti-Nembe et al. 2013).

Application of Irradiation with Edible Coating

Irradiation combined with hot-water immersion or calcium ascorbate treatment was shown to be effective for maintaining the postharvest quality of apples and papayas (Rashid et al. 2015). In addition to fumigation treatment, apples exposed to low doses of gamma radiation (<0.4 kGy) showed increased disease resistance. Combined treatments reduced the necessary irradiation dose in phytosanitary irradiation processing under storage conditions (Cheon et al. 2016b). Coating a fruit decreased its weight loss during storage by providing a water and gas barrier (Valencia-Chamorro et al. 2010). On the contrary, irradiation maintained the fruit weight by slowing metabolic processes. Gamma irradiation (0.5 kGy), combined with commercial edible coating (Sta-Fresh 2505), was shown to lower the weight loss in golden-yellow and purple-red tamarillo (*Solanum betaceum* Cav.) fruits by 19%; coating reduced it by 18%, while use of treatments in combination reduced it by 37% (Abad et al. 2017). Polyphenol-chitosan coating can increase the antimicrobial effect on irradiation. Total viable counts and psychrophiles in samples subjected to the combined treatment were lower than those in samples treated with only chitosan coating or irradiation (Zhang et al. 2017). The use of bioactive coating under modified atmosphere packaging synergistically increased the radiosensitization of *E. coli* and *S. typhimurium* in food by 1.80- and 1.89-fold, respectively (Severino et al. 2015).

Gamma ray, electron beam and X-ray are effective for sanitizing food products according to the international standards proposed for export quality foods. Sterilization using ionizing radiations has been studied extensively and approved by private and governmental organizations. The antimicrobial effect without substantial increase in temperature can be direct or indirect. The direct effect is achieved by damaging the microbial DNA in the treated food products by ionizing particles or rays, which prevents microbial proliferation due to inhibition of DNA synthesis. The indirect effect stems from the interaction with water molecules, which produces hydroxyl (OH^-) and hydrogen (H^+) radicals and hydrated electrons ($e^-_{(aq)}$) that together cause cell lysis (Mohácsi-Farkas et al. 2014). Direct effects are typical in low moisture foods, where the dose response is used to understand microbial inactivation kinetics, while indirect outcome is observed in foods with high moisture content. There are geographical differences in the preferred methods: for instance, North America prefers UV and other ionizing radiations, while European countries find PEF as the better choice. Gamma ray, X-ray, and electron beam applications scored high on fresh produce (71%), meat, and poultry (67%), as ionizing radiation is more suitable for solid products. Irradiation has been the well-known approved application in a recent survey for use in raw meat, eggs, fruits, vegetables, and spices (Jermann et al. 2015). The efficacy of its application can also be extended to other products, such as fish and oysters (Praveen et al. 2013). Combination of these technologies improves the efficiency of decontamination and overcomes the limitations of a specific technology.

Radiation-Induced Changes in Pre-Packaged Foods, Effects on Barrier Properties of the Packaging Material, Technical Challenges, and Safety Issues

Nicholas Appert invented the caning process in 1810, thus triggering the onset of modern food packaging era. Plastic was introduced as a packaging material in 1930s, and its use has increased thereafter (Robertson, 2016). The global market of food packaging is estimated to be an industry worth of US $440 billion, increasing at the rate of 10% annually. Approximately 65% of the total packaging material produced is employed by the food industry alone. Polymers from the petrochemical industry were mainly used for manufacturing of packaging materials. The most commonly-used plastic materials used in the food industry are as follows: polypropylene (PP), polyethylene (PE), poly (ethylene terephthalate) (PET), polystyrene (PS), and poly (lactic acid) (PLA). These materials are preferred due to their versatile properties, such as ease of recycling, mechanical strength, resistance to thermal extremes, good barrier characteristics against volatiles and gaseous exchange (Yam and Lee, 2012).

Recently, the trend in the food industry has shifted from traditional metal, glass, and paper towards polymeric materials. Polymer-based packaging materials are treated with ionizing radiations for achieving improved glaze-resistance, barrier function, adhesiveness, printability, and antimicrobial properties for enhanced food safety (Komolprasert, 2016). Irradiation of pre-packaged food materials using gamma or e-beam may be employed for ensuring food safety and preservation due to improved barrier properties. Packaging materials may undergo several modifications on exposure to radiation treatment. These modifications may affect their barrier properties, thus ultimately influencing the functionality and resistance of these materials against chemicals and microbes (Madera-Santana et al. 2016). Moreover, different food additives, such as stabilizers and antioxidants, are sensitive to radiation-induced radiolytic products, and may degrade or breakdown when polymeric materials are exposed to ionizing radiations. The degradation may occur preferentially on the surface of the irradiated materials, potentially resulting in transfer of generated radiolytic products into the food products (Komolprasert, 2014; Komolprasert and Turowski, 2014). Packaging materials made of paper, paperboard, cardboard, brick carton, and glass do not exhibit any potential hazards on exposure to ionizing radiation sources. However, plastic and polymeric packaging materials may undergo several chemical changes after exposure to ionizing radiation, compromising the safety of the pre-packaged foods. Therefore, it is necessary to evaluate the effects of ionizing radiations on polymeric packaging materials, as discussed below.

Effects of Gamma Irradiation on Packaging Materials

Gamma irradiation from Cobalt-60 source is a well-known cold sterilization technique that is non-polluting, is effective at ambient temperature, and offers great degree of flexibility because of process control (Madera-Santana et al. 2016). Furthermore, gamma irradiation treatment is not just limited to superficial context, as penetration depth is high enough in case of ionizing radiation after exposure to photon energy. Exposure under experimental conditions has revealed that polymeric packaging materials usually undergo superficial changes without any significant effect on the food commodities packed inside (Silvestre and Cimmino, 2013). The main functionalization advantage offered by gamma irradiation is attachment of the desired functional group on the surface of the polymeric packaging material. These attached functional groups may enhance the barrier properties of the packaging material by immobilizing enzymes and other bioactive components. Polymeric materials, PE, PS, and PP were processed for radiation-induced grafting of acrylic acid on the surface. Enhanced yields of radicals and hydrogen emission were reported in PS, followed by PP and PE, as a result of exposure to gamma irradiation. Gamma irradiation improved the sterilization effect with increased attachment of functional groups to facilitate immobilization of bioactive components (Przybytniak et al. 2008). PLA is another famous bio-degradable packaging material known for its transparency, plasticity, and higher degree of mechanical strength, and it is also approved for rigid and flexible packaging purpose on commercial scale by US Food and Drug Administration (FDA, 2010). High-dose gamma irradiation treatment at > 60 kGy induced several changes in the PLA matrix. PLA matrix showed a dose-dependent decrease in tensile strength, and melting temperatures decreased, while optical properties, such as colour difference and opacity showed an increasing tendency with increase in the applied irradiation dose. The average molecular weight decreased as a result of oxidative degradation and chain scissioning of PLA clamshell samples at higher doses (Madera-Santana et al. 2016). Similarly, feasibility of gamma irradiation for preparing chitosan films grafted by monomer, 2-hydroxyethyl methacrylate (HEMA) for food packaging applications was evaluated. The chitosan solutions were exposed to low doses of gamma irradiation (0.1–0.3 kGy), and chitosan films were further grafted by HEMA. The gamma-irradiated HEMA–grafted chitosan performed better in terms of mechanical and barrier properties, as compared to the non-irradiated control samples, and the authors concluded that HEMA–grafted chitosan films might be used as bio-degradable food packaging materials with enhanced antimicrobial properties to improve the preservation status of various pre-packaged foods (Khan et al. 2012). The gamma-irradiated packaging materials were studied for monitoring changes in migration rates of volatile compounds. Gamma irradiation caused a significant reduction in the migration of volatile compounds, as compared to that observed in non-irradiated materials, under the threshold limit established by European Union (EU) regulation (10/2011) during 8 months of storage period without any significant changes with increasing applied gamma irradiation doses (Salafranca et al. 2015). Similarly, PE-based materials (with low and high densities) were exposed to gamma irradiation at dose range of 5–60 kGy. Gamma irradiation treatment resulted in the production of a large number of radiolytic products, which increase dose-dependently. High dose treatment with 30 and 60 kGy led to slight modifications in the radiolytic products, which could be explained in terms of radiation-induced oxidation of PE-based materials (Chytiri et al. 2008).

Effects of E-beam and Pulsed X-Rays Irradiation on Packaging Materials and Their Barrier Properties

E-beam irradiation has also been used for imparting desirable changes in packaging materials. Polymeric plastic packaging materials are usually exposed to e-beam irradiation in the dose range of $\leq$ 1 kGy or below 10 kGy. The e-beam treatment gained popularity due to its peculiar features of absence of radioactive isotopes, rapid results, and cost-effectiveness of the process by generation of high energy electrons after electric ionization (Pillai and Shayanfar, 2014). Exposure to e-beam irradiation leads to the following major changes: i) bond cleavage due to absorption of electrons and subsequent generation of free radicals, followed by formation of crosslinks and end-links; ii) recombination of generated radials, which give rise to gas evolution and chain scissioning, also known as disproportionation (Salvatore et al. 2016); and iii) death of microorganisms for sterilization purpose. E-beam has been used in recent years for improvement of the barrier properties of polymeric and flexible packaging materials, as their use is rising steadily on account of their enhanced protective properties, convenience, and traceability (Salvatore et al. 2016). Regarding the effects of e-beam irradiation on polymeric packaging materials, it was reported that e-beam exposure led to formation of new chromophore groups that efficiently absorbed the applied electromagnetic radiations from various spectral regions (UV, IR, and visible). Moreover, e-beam also causes crosslinking and degradation of bonds in the main polymer chain, followed by cyclic generation of intermolecular bonds and gaseous moieties, such as methane, hydrogen, and carbon monoxide. These phenomena also depend on the applied e-beam dose and oxygen content in the surrounding environment. Moreover, chemical processes involve thermal treatment of polymeric packaging materials, which may produce unwanted degradation products and polymer melting. On the contrary, e-beam irradiation, as a non-thermal treatment, offers pertinent advantage of imparting *in situ* desirable changes in packaging materials (crosslinking and end-linking) in the finished solid-state form at low temperatures without any need of melting or heating the polymeric material. E-beam is also reported to enhance the adhesiveness and flexibility of the packaging materials (Uchida, 2008). Polymeric packaging films were treated with e-beam irradiation to improve the oxygen barrier properties for shelf-life extension. Spoilage of oxygen-sensitive food products may be delayed by limiting the oxygen exposure, and improved oxygen barrier function may result in reduced inventory costs. Packaging materials are processed to provide an active oxygen barrier, which offers a uniform scavenging effect by decreasing the oxygen levels throughout the packaged foods. An e-beam dose of 2 kGy has been reported to improve the scavenging abilities, also known as active oxygen barrier properties, and e-beam irradiated films showed no or low oxygen absorption, whereas the oxygen ingress rate in non-irradiated samples was steady even after 6 months of storage. As a result, this multilayer film could be used for improved packaging applications with minimum trapped oxygen inside the food packages (Beckwith et al. 2016).

Pulsed X-ray is another ionizing radiation that involves application of high voltage pulses through food products at refrigerated or ambient temperatures. Pulsed X-rays are usually supplied by the electricity-driven source, and usually offer ease of incorporation into the existing operational system, as a non-thermal irradiation treatment. As radionucleotides are not involved, permanent massive shielding, as required by gamma irradiation facility, is not necessary. Instead, high intensity pulsed X-rays (up to 5 MeV) are generated by employing a solid-state switch panel. At present, the X-ray treatment has gained popularity because it allows orientation of the radiation field to accommodate food packages of variable sizes with high degree of versatility and reproducibility (Morris et al. 2007). Exposing polymeric packaging materials to these high intensity pulsed X-rays may lead to physical and chemical alterations of these polymeric structures (Mahalik and Nambiar, 2010). Similarly, e-beam and gamma irradiation, and radiation processing by employing X-rays may also result in significant structural and conformational modifications of materials, including formation of reactive moieties and free radicals, bond cleavage with new bonding arrangement, scissioning of chain length (degradation followed by decreased molecular weight), crosslinking (increase in molecular weight), and end-linking (Haji-Saeid et al. 2007). As per the approved FDA regulations, the following materials can be safely exposed to irradiation treatment up to 10 kGy without any significant effects on their barrier properties: coated cellulose film, glassine paper, waxed paperboard, kraft paper, PET and PS. Among durable and resistant polymers, vegetable parchment, PE films, PET, nylon 6, and acrylonitrile copolymers are safe to be irradiated with dosages up to 60 kGy (Machi, 2004; Nielsen and Jägerstad, 1994).

Technical Challenges and Consumer Safety Issues

Irradiation of polymeric plastic materials may result in generation of free radicals, followed by chain scission and crosslinking phenomena. Consequently, volatile radiolytic products are produced, which may compromise the migration properties of packaging materials. Ionizing radiation treatments significantly influence the migration characteristics of additives and additive-containing decomposition products. Changes in both sensory and migration properties may exhibit strong effects on quality characteristics of pre-packaged foods, consequently affecting consumer safety (Arvanitoyannis, 2010).

Prior to the irradiation treatment, foodstuffs are usually packaged into the final form. This reduces the likelihood of possible microbial contamination in foodstuffs after exposure to ionizing radiation. It also allows simultaneous irradiation of packaging materials along with the food itself. However, care should be taken that irradiated packaging materials are not be modified in a manner that may lead to migration of harmful chemical compounds into the packaged foodstuffs. In the past, FDA approved several types of materials for packaging purposes, and new packaging materials with versatile oxygen barrier properties have also been introduced in the recent years.

The safety of these newly-introduced materials prior to their distribution in the market is of paramount importance to the food industry and regulatory agencies like FDA, keeping in view the effects of irradiation sources and compliance with the regulatory requirements (Komolprasert, 2016). Federal, Food, Drug, and Cosmetic Act (section 201), 21 Code of Federal Regulations (CFR) 179, is the primary FDA regulation that deals with the safety assessment of food packaging materials exposed to ionizing radiations (gamma, e-beam, X-rays). Most importantly, this regulation specifies the mandatory pre-market approval of packaging materials before production on a commercial scale. Most of the approved packaging materials by FDA for pre-packaged foods are allowed to be irradiated up to 10 kGy dose, irrespective of the irradiation type, as gamma, e-beam and X-ray were concluded to be equivalent in terms of generation of radiolytic products. In accordance with the most recent regulation under CFR 170.39, pre-packaged foods should be irradiated under an inert environment/vacuum or under frozen conditions with essentially zero level of generated radiolytic products (Komolprasert, 2016).

Radiation–induced crosslinking and degradation of polymeric packaging materials may lead to chemical changes in the materials. In an inert or vacuum environment, by complying with the FDA regulation, the cross-linking (polymerization) takes precedence over degradation, whereas irradiation in presence of air or oxygen may result in domination of degradation over polymerization, due to chain scissioning (Komolprasert,

2014). Irradiation under inert, frozen, or vacuum atmosphere is thus an important criterion under CFR 170.39 for granting the exemption request to the packaging materials that are irradiated by being in contact with the foodstuffs.

Additives, such as stabilizers or antioxidants, along with base polymeric materials might be subjected to degradation process induced by ionizing radiation. As a result of this degradation, generated radiolytic products may migrate into the foodstuffs packed inside, and organoleptic characteristics and safety of pre-packaged foods could be compromised. Therefore, premarket assessment is mandatory to ensure the safety status of new packaging materials, and pre-packaged foods exposed to high dose of irradiation levels under presence of oxygen must be evaluated via migration test prior to their intended use (Komolprasert and Turowski, 2014; FDA, 2010).

Moreover, the quantitative and qualitative analyses of the migrants could be a useful approach for identification of the harmful radiolytic products in foodstuffs in contact with irradiated packaging materials and to evaluate the effects of ionizing radiation on the barrier properties. The major concern in this regard is identification of unknown migrants and assessment of the toxicological effects of radiolytic products at the recommended dietary concentration of more than 0.5 parts per billion. As most of the radiolytic compounds from irradiated packaging materials resemble the industrial chemicals in terms of their structural configuration, the structure-activity relationship (SAR) was recommended as a useful analytical method by Van Bossuyt et al. (2017) for food components with toxicological concerns arising from irradiated packaging materials.

Application of Electronic Sensing Techniques to Discriminate Irradiated Foods

Screening of irradiated foods is very important to regulate the trade of irradiated foods on a global market. In this regard, several screening and detection methods have been documented in the past few decades: photostimulated luminescence (EN13751, 2009), thermoluminescence (EN1788, 2001), electron spin resonance spectroscopy (EN13708, 2001), direct epifluorescent filter technique and aerobic plate count (EN13783, 2001), and DNA comet assay (EN13874, 2001). Recently, electronic sensing techniques, such as E-nose and E-tongue have emerged as new screening tools to differentiate the irradiated foods from non-irradiated counterparts on the basis of volatile flavour profiles and unique taste attributes (An et al. 2017).

Taste and smell are the primary attributes that give rise to olfaction properties in combination. In the recent years, electronic sensing techniques, such as E-nose and E-tongue, have emerged as powerful and widely used instrumental tools for mimicking the human tasting and smelling abilities (Qiu et al. 2014).

The human olfaction system exhibits high complexity owing to thousands of sensory receptors with specificity towards various types of taste and aromas (Aprotosoaie et al. 2016; Du et al. 2010). Odour molecules from various foodstuffs can bind to these sensory receptors (ranging from 10–100 million), allowing detection of odours up to part per billion level. The human brain classifies the signals detected by the sensory receptors, and foodstuffs are categorized with respect to their aromatic and taste profiles. E-nose and E-tongue applications were brought into light for the first time as a non-destructive technique during 1980s, and developed instruments were introduced for commercial purposes in the year 1993 (Majchrzak et al. 2017; Brudzewski et al. 2010).

E-nose comprises of an array of different chemical sensors in its configuration, and these sensors detect the changes in volatiles flavour patterns in an electrochemical manner. These signals were processed by the digital patterns-recognition algorithms, and different food components were classified by discrimination of diverse mixtures of possibly existing analytes. Individual sensors from both E-nose and E-tongue generate distinctive patterns in the form of a digital response (Qiu et al. 2014). The applicability of E-sensing techniques for irradiated foods, determined by principal component analysis, has been shown in Figure 1.

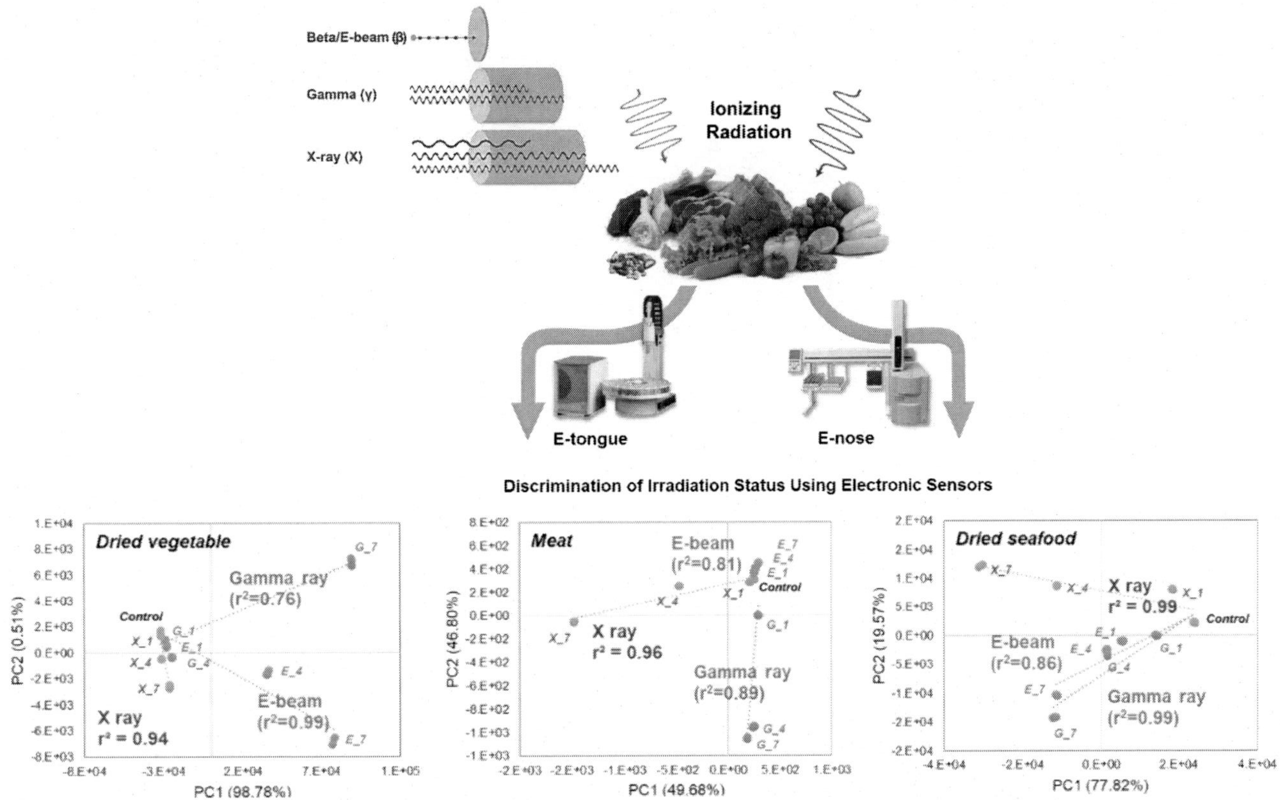

Figure 1. Electronic sensing techniques (E-nose and E-tongue) for discriminating irradiated foods and elucidation by principal component analysis.

Among organoleptic features, taste is recognized as an important parameter during mastication process owing to the presence of taste receptors in the oral cavity of human beings. Different components of organic and inorganic origin interactively produce unique taste attributes of foodstuffs. The E-tongue sensor arrays consist of various electrochemical sensors, which detect these taste attributes in a similar manner to that of the human gustation system (Wesoły et al. 2016; Wilson and Baietto, 2009).

E-tongue is an instrumental technique that can mimic the flavouring and smelling abilities of human olfactory receptors in rapid and simple manner for assessment of the volatile flavour profiles of wide variety of foodstuffs. E-nose interacts with the volatile flavour molecules in the food matrix and provides valuable information with respect to compositional modifications of foodstuffs after exposure to ionizing radiations, and such evaluations are beyond the scope of human sensory perceptions and compromise the results of sensory evaluation by trained panelists. On the basis of these signals, further classification of foodstuffs is performed with the help of pattern recognitions (Santos et al. 2010). Overall, the e-sensing techniques like E-nose and E-tongue have proven to be successful instrumental tools for assessing the properties of foods, detecting food adulteration, and classifying foods of diverse origins based on their volatile flavour and taste profiles. Currently, E-nose and E-tongue have been used in combination with gas chromatography to analyse volatile flavours and taste patterns both qualitatively and quantitatively. Recently, E-sensing techniques have gained popularity for discriminating the irradiated food from non-irradiated control samples.

Virgin coconut oil was gamma-irradiated by Ghosh et al. (2016) to remove the rancid odour from expeller-pressed oil. Coconut oil samples irradiated at 4.2 kGy showed no rancid odour, and E-nose analysis was used for confirmation of absence of the rancid odour after a storage period of 28 days. Moreover, the authors reported that gamma irradiation did not impart any effect on the antioxidant potential of coconut oil (Ghosh et al. 2016). In another study, changes in E-beam irradiated cumin and chili powders at doses of 6, 10 and 14 kGy were investigated by Sanyal et al. (2014). PCA confirmed the distinct groupings for both irradiated and control samples depending on the applied irradiation doses. As compared to cumin, clearly visible distinct groups were observed in case of irradiated chili powder samples, as compared with the non-irradiated control. The authors have described E-nose as a potentially feasible screening tool to discriminate irradiated samples.

As a novel screening approach, E-nose analysis was used to evaluate the effects of gamma irradiation on flavour and odour properties of Korean red pepper powder samples (Lee et al. 2004). Microbial loads are usually higher in case of spice samples, and thus microbial decontamination is necessary. This is usually accomplished by treating spices with ionizing radiations. The Korean red pepper powders were exposed to gamma irradiation up to 7 kGy in vacuum packed form. Metal oxide sensors (Ghosh et al. 2016) were employed for detecting changes in the irradiated spice samples, and modifications were recorded as a measure of change in resistivity of the array sensors. This change in resistivity yielded to the E-nose fingerprints from all the samples (0, 3, 5 and 7 kGy). These E-nose fingerprints were utilized for elucidating the numerical data trends by principal component analysis (PCA). This multivariate analysis of the irradiated and non-irradiated control samples clearly distinguished the irradiated Korean spice samples from the non-irradiated counterparts, based on volatile odour profiles. It was suggested by these findings that gamma irradiation treatment significantly ($p < 0.05$) influenced the volatile flavour profiles, and E-nose could serve as the potential screening tool for irradiated spice samples (Lee et al. 2004).

Similarly, salted and fermented anchovy was subjected to gamma irradiation at 0, 2.5, 5, 7.5, and 10 kGy. The irradiated and non-irradiated samples were evaluated for changes in their odour patterns by E-nose analysis during storage. Samples irradiated at 5 and 10 kGy demonstrated different odour patterns in comparison with the control. The irradiated samples were distinguished on the basis of dose-dependent variations (Kim et al. 2004). Akram et al. (2012) have also reported the possible feasibility of E-nose as a screening tool to identify irradiated mushrooms (*Pleurotus eryngii*) when exposed to gamma irradiation at 0, 1, 2, and 3 kGy doses.

The PCA score plot demonstrated clearly visible groups for both irradiated and control samples on the basis of changes in the volatile flavour patterns. E-nose has also been utilized by An et al. (2017) for discrimination of e-beam-irradiated smoked duck meat samples. The authors have reported that PCA of E-nose data demonstrated successful application of E-nose to detect volatile flavour changes in irradiated smoked duck meat during different storage periods (0–40 days) (An et al. 2017). In summary, the E-nose and E-tongue analyses exhibit potential to discriminate irradiated foods based on screening of volatile flavour patterns and unique taste attributes. However, it must be noted that E-sensing techniques are merely screening tools for irradiated foodstuffs, as changes in volatile flavour patterns are reportedly not radiation-specific detection markers.

REFERENCES

Abad, J., Valencia-Chamorro, S., Castro, A., Vasco, C., 2017. Studying the effect of combining two nonconventional treatments, gamma irradiation and the application of an edible coating, on the postharvest quality of tamarillo (*Solanum betaceum* Cav.) fruits. *Food Control* 72, 319-323.

Abboudi, M., Al-Bachir, M., Koudsi, Y., Jouhara, H., 2016. Combined effects of gamma irradiation and blanching process on acrylamide content in fried potato strips. *Int. J. Food Properties*, 19, 1447-1454.

Ahn, J. J., Akram, K., Jeong, M. S., Kwak, J. Y., Kwon, J. H., 2013. Identification of irradiated shellfish using well-characterized thermoluminescence properties of biogenic minerals present in the exoskeletons. *Food Anal. Met.* 6, 1345-1352.

Ahn, J. J., Sanyal, B., Park, E. J., Lim, H. K., Kwon, J. H., 2014. Effects of sample pretreatments on EPR spectral characteristics of irradiated sea algae–an advanced approach to identify irradiation status. *RSC Adv.* 4, 32312-32319.

Akram, K., Ahn, J. J., Yoon, S. R., Kim, G. R., Kwon, J. H., 2012. Quality attributes of *Pleurotus eryngii* following gamma irradiation. *Postharvest Biol. Technol.* 66, 42-47.

Alahakoon, A. U., Jayasena, D. D., Jung, S., Kim, S. H., Kim, H. J., Jo, C., 2015. Effect of electron beam irradiation and high pressure treatment combined with citrus peel extract on seasoned chicken breast meat. *J. Food Process. Preserv.* 39, 2332-2339.

An, K. A., Arshad, M. S., Jo, Y., Chung, N., Kwon, J. H., 2017. E-Beam irradiation for improving the microbiological quality of smoked duck meat with minimum effects on physicochemical properties during storage. *J Food Sci.* 82, 865-872.

Aprotosoaie, A. C., Luca, S. V., Miron, A., 2016. Flavor chemistry of cocoa and cocoa products – an overview. *Compr. Rev. Food Sci. Food Saf.* 15, 73-91.

Arshad, K., Sudha, K., Hatha, A. A. M., Anilkumar, G., 2015. Combined effect of gamma irradiation and cold temperature storage on the sensory properties of edible estuarine crab *Scylla serrata*. *Int. Food Res. J.* 22, 2253-2258.

Arvanitoyannis, I. S., 2010. *Irradiation of food commodities: Techniques, applications, detection, legislation, safety and consumer opinion.* London: Academic Press, pp. 43.

Aylangan, A., Ic, E., Ozyardimci, B., 2017. Investigation of gamma irradiation and storage period effects on the nutritional and sensory quality of chickpeas, kidney beans and green lentils. *Food Control* 80, 428-434.

Banerjee, R., Jayathilakan, K., Chauhan, O. P., Naveena, B. M., Devatkal, S., Kulkarni, V. V., 2017. Vacuum packaged mutton patties: Comparative effects of high pressure processing and irradiation. *J. Food Process. Preserv.* 41, 1-9.

Bárdos, L., Baránková, H., 2010. Cold atmospheric plasma: Sources, processes, and applications. *Thin Solid Films* 518, 6705-6713.

Beckwith, S., Edwards, F. B., Rivett, J., Ebner, C. L., Kennedy, T., McDowell, R., Speer, D. V., 2016. U.S. Patent No. 9,452,592. Washington, DC: U.S. Patent and Trademark Office.

Bhat, N. A., Wani, I. A., Hamdani, A. M., Gani, A., Masoodi, F. A., 2016. Physicochemical properties of whole wheat flour as affected by gamma irradiation. *LWT-Food Sci. Technol.* 71, 175-183.

Blessington, T., Scheuring, D. C., Nzaramba, M. N., Hale, A. L., Reddivari, L., Vestal, T. A., Maxim, J. E., Miller, J. C., 2015. The use of low-dose electron-beam irradiation and storage conditions for sprout control and their effects on xanthophylls, antioxidant capacity, and phenolics in the potato cultivar Atlantic. *Am. J. Potato Res.* 92, 609-618.

Brudzewski, K., Osowski, S., Ulaczyk, J., 2010. Differential electronic nose of two chemo sensor arrays for odor discrimination. *Sens. Actuators B* 145, 246-249.

Cetinkaya, N., Ercin, D., Özvatan, S., Erel, Y., 2016. Quantification of applied dose in irradiated citrus fruits by DNA Comet Assay together with image analysis. *Food Chem. 192*, 370-373.

Cheon, W., Kim, Y. S., Balaraju, K., Kim, B. S., Lee, B. H., Jeon, Y., 2016a. Postharvest control of *Botrytis cinerea* and *Monilinia fructigena* in apples by gamma irradiation combined with fumigation. *J. Food Prot. 79*(8), 1410-1417.

Cheon, W., Kim, Y. S., Balaraju, K., Kim, B. S., Lee, B. H., Jeon, Y., 2016b. Postharvest disease control of *Colletotrichum gloeosporioides* and *Penicillium expansum* on stored apples by gamma irradiation combined with fumigation. *Plant Pathol. J.* 32, 460-468.

Chowdhury, M. A. Z., Jahan, I., Karim, N., Alam, M. K., Rahman, M. A., Moniruzzaman, M., Gan, S. H., Fakhruddin, A. N. M., 2014. Determination of carbamate and organophosphorus pesticides in vegetable samples and the efficiency of gamma-radiation in their removal. *BioMed Res. Int.* 2014, 1-9.

Chytiri, S., Goulas, A. E., Badeka, A., Riganakos, K. A., Petridis, D., Kontominas, M. G., 2008. Determination of radiolysis products in gamma-irradiated multilayer barrier food packaging films containing a middle layer of recycled LDPE. *Radiat. Phys. Chem.* 77, 1039-1045.

Domijan, A. M., Pleadin, J., Mihaljević, B., Vahčić, N., Frece, J., Markov, K., 2015. Reduction of ochratoxin A in dry-cured meat products using gamma-irradiation. *Food Addit. Contam. Part A* 32, 1185-1191.

Du, X., Bai, J., Plotto, A., Baldwin, E., Whitaker, V., Rouseff, R., 2010. Electronic nose for detecting strawberry fruit maturity. *Proc. Fla. State Hort.* 123, 259-263.

El-Nesr, E. M., Raafat, A. I., Nasef, S. M., Soliman, E. A., Hegazy, E. A., 2013. Chitin and chitosan extracted from irradiated and non-irradiated shrimp wastes (Comparative analysis study). *Arab J. Nuclear Sci. Appl.* 46, 53-66.

EN1788, 2001. *Foodstuffs-Thermoluminescence detection of irradiated food from which silicate minerals can be isolated.* Brussels, Belgium: European Committee of Standardization (CEN).

EN13708, 2001. *Foodstuffs-detection of irradiated food containing crystalline sugar by ESR spectroscopy*. Brussels, Belgium: European Committee of Standardization (CEN).

EN13751, 2009. *Foodstuffs-Detection of irradiated food using photostimulated luminescence*. Brussels, Belgium: European Committee of Standardization (CEN).

EN13783, 2001. *Foodstuffs-Detection of irradiated food using direct epifluorescent filter technique/aerobic plate count (DEFT/APC)-screening method.* Brussels, Belgium: European Committee of Standardization (CEN).

EN13874, 2001. *Foodstuffs-DNA comet assay for the detection of irradiated foodstuffs-screening method.* Brussels, Belgium: European Committee of Standardization (CEN).

Eustice, R., 2011. *Food irradiation: A global perspective and future prospects*. http://ansnuclearcafe.org/2011/06/09/food-irradiation-a-global-perspective-future-prospects/.

Fadhel, Y. B., Leroy, V., Dussault, D., St-Yves, F., Lauzon, M., Salmieri, S., Jamshidian, M., KhanhVu, D., Lacroix, M., 2016. Combined effects of marinating and γ-irradiation in ensuring safety, protection of nutritional value and increase in shelf-life of ready-to-cook meat for immunocompromised patients. *Meat Sci.* 118, 43-51.

Fanaro, G. B., Hassimotto, N. M. A., Bastos, D. H. M., Villavicencio, A. L. C. H., 2014. Effects of γ-radiation on microbial load and antioxidant proprieties in black tea irradiated with different water activities. *Radiat Phys. Chem.* 97, 217-222.

Fanaro, G. B., Hassimotto, N. M. A., Bastos, D. H. M., Villavicencio, A. L. C. H., 2015. Effects of γ-radiation on microbial load and antioxidant proprieties in green tea irradiated with different water activities. *Radiat. Phys. Chem.* 107, 40-46.

FDA, 2010. *Threshold of regulation exemptions.* http://www.fda.gov/Food/IngredientsPackagingLabeling/PackagingFCS/ThresholdRegulationExemptions/ucm093685.htm.

Feng, X., Ahn, D. U., 2016. Volatile profile, lipid oxidation and protein oxidation of irradiated ready-to-eat cured turkey meat products. *Radiat. Phys. Chem.* 127, 27-33.

Feng, X., Moon, S., Lee, H., Ahn, D. U. 2016. Effect of irradiation on the parameters that influence quality characteristics of uncured and cured cooked turkey meat products. *Poultry Sci.* 95(12), 2986-2992.

Fernandes, Â., Barreira, J. C., Antonio, A. L., Rafalski, A., Morales, P., Férnandez-Ruiz, V., Oliveira, M. B. P. P., Martins, A., Ferreira, I. C., 2016. Gamma and electron-beam irradiation as viable technologies for wild mushrooms conservation: Effects on macro-and micro-elements. *Eur. Food Res. Technol.* 242, 1169-1175.

Fernandes, Â., Barreira, J. C., Günaydi, T., Alkan, H., Antonio, A. L., Oliveira, M. B. P., Martins, A., Ferreira, I. C., 2017. Effect of gamma irradiation and extended storage on selected chemical constituents and antioxidant activities of sliced mushroom. *Food Control* 72, 328-337.

Finten, G., Garrido, J. I., Agüero, M. V., Jagus, R. J. 2017. Irradiated ready-to-eat spinach leaves: How information influences awareness towards irradiation treatment and consumer's purchase intention. *Radiat. Phys. Chem. 130*, 247-251.

Gayán, E., Serrano, M. J., Raso, J., Álvarez, I., Condón, S., 2012. Inactivation of *Salmonella enterica* by UV-C light alone and in combination with mild temperatures. *Appl. Environ. Microbiol.* 78, 8353-8361.

Ghosh, P. K., Chatterjee, S., Bhattacharjee, P., Bhattacharyya, N., 2016. Removal of rancid-acid odor of expeller-pressed virgin coconut oil by gamma irradiation: Evaluation by sensory and electronic nose technology. *Food Bioprocess Technol.* 9, 1724-1734.

Ha, J. W., Kang, D. H., 2015. Enhanced inactivation of food-borne pathogens in ready-to-eat sliced ham by near-infrared heating combined with UV-C irradiation and mechanism of the synergistic bactericidal action. *Appl. Environ. Microbiol.* 81, 2-8.

Haji-Saeid, M., Sampa, M. H. O., Chmielewski, A. G., 2007. Radiation treatment for sterilization of packaging materials. *Radiat. Phys. Chem.* 76, 1535-1541.

Hallman, G. J. 2017. Process control in phytosanitary irradiation of fresh fruits and vegetables as a model for other phytosanitary treatment processes. *Food Control 72*, 372-377.

Hong, Y. S., Kim, K. S., 2014. Comparison of radiation-induced hydrocarbons in dried anchovies by electron-beam irradiation. *J. Korean Soc. Food Sci. Nutr.* 43, 1889-1895.

Huang, S. J., Liu, Y. L., Mau, J. L., Tsai, S. Y., 2014. The content of physiologically active substances in several medicinal mushrooms with various doses of γ-irradiation. In: Advanced materials research. *Trans Tech Publications*, pp. 707-711.

Huq, T., Vu, K. D., Riedl, B., Bouchard, J., Lacroix, M., 2015. Synergistic effect of gamma (γ)-irradiation and microencapsulated antimicrobials against *Listeria monocytogenes* on ready-to-eat (RTE) meat. *Food Microbiol.* 46, 507-514.

Ichinose, K., Yasuda, K., Yamashita, N., Matsumura, M., Okada, Y., 2016. Reduced dispersal and survival in the sweet potato weevil (*Euscepes postfasciatus*) after irradiation. *Agric. Forest Entomol.* 18, 157-166.

Ihsanullah, I., Rashid, A., 2017. Current activities in food irradiation as a sanitary and phytosanitary treatment in the Asia and the Pacific region and a comparison with advanced countries. *Food Control* 72, 345-359.

Janagam, V. M., Sethi, S., Kaur, C., Singh, B., Pal, R. K., 2014. "Determination of effective irradiation dose to maintain quality of garlic cultivars during storage". In: *Proceedings of the International*

Conference on Radiation Biology: Frontiers in Radiobiology-Immunomodulation, Countermeasures and Therapeutics: Abstract Book, Souvenir and Scientific Programme.

Jeong, I. S., Khan, N., Jamila, N., Kim, K. S., 2017. Determination of irradiation induced hydrocarbons in beef jerky by GC-MS. *J. Chem. Soc. Pakistan* 39, 119-124.

Jermann, C., Koutchma, T., Margas, E., Leadley, C., Ros-Polski, V., 2015. Mapping trends in novel and emerging food processing technologies around the world. *Innovative Food Sci. Emerging Technol.* 31, 14-27.

Jo, D., Sanyal, B., Lee, J. W., Kwon, J. H. 2015. A thermoluminescence study to identify irradiation of imported oranges exposed to electron, X-ray and gamma radiation for quarantine application. *J. Radioanal Nucl. Chem. 303*, 297-304.

Jo, Y., Sanyal, B., Park, H. J., Kwon, J. H. 2016. Ethanol extraction-based drying enhanced ESR radical detection in oranges irradiated to different ionizing radiations during storage. *Postharvest Biol. Tec. 112*, 170-175.

Jo, Y., Sanyal, B., Ameer, K., & Kwon, J. H. 2018. Characterization of DNA comet and cellulose radical signal in Valencia oranges treated with different forms of ionizing radiation. *Postharvest Biol. Tec. 135*, 68-76.

Jouki, M., Khazaei, N. 2014. Effect of low-dose gamma radiation and active equilibrium modified atmosphere packaging on shelf life extension of fresh strawberry fruits. *Food Packag. Shelf Life* 1(1), 49-55.

Jung, D. W., Huang, Y. H., Song, B. S., Byun, M. W., Kang, I. J., 2014. Genotoxicological safety evaluation of x-ray irradiated four foods. *J. Korean Soc. Food Sci. Nutr.* 43, 1588-1593.

Kancheva, V. D., Slavova-Kazakova, A., Terzieva, A., Ivanova, M., Tsrunchev, T., 2015. Assessing the potential of some traditional Bulgarian teas in scavenging free radicals and their antioxidant activity after gamma-irradiation. *Riv. Ital. Delle Sostanze Grasse* 92, 175-182.

Kang, S., Park, S. Y., Ha, S. D., 2016. Application of gamma irradiation for the reduction of norovirus in traditional Korean half-dried seafood products during storage. *LWT-Food Sci. Technol.* 65, 739-745.

Kang, J. A., Song, H. Y., Byun, E. H., Ahn, N. G., Kim, H. M., Nam, Y. R., Lee, G. H., Jang, B. S., Choi, D. S., Lee, D. E., Byun, E. B., 2017. Gamma-irradiated black ginseng extract inhibits mast cell degranulation and suppresses atopic dermatitis-like skin lesions in mice. *Food Chem. Toxicol.* 111, 133-143.

Khalil, A., Al-Bachir, M., 2015. The deployment of γ-irradiation for reducing polycyclic aromatic hydrocarbons and microbial load in wheat kernels. *Toxicol. Environ. Chem.* 97, 857-867.

Khan, A., Huq, T., Khan, R. A., Dussault, D., Salmieri, S., Lacroix, M., 2012. Effect of gamma radiation on the mechanical and barrier properties of HEMA grafted chitosan-based films. *Radiat. Phys. Chem.* 81, 941-944.

Khan, N., Jamila, N., Choi, J. Y., Nho, E. Y., Imran, M., Hussain, I., Kim, K. S., 2017. Effect of electron beam irradiation on the volatile flavor profile of *Elettaria cardamomum* (l.) Maton., from Pakistan. *J. Chem. Soc. Pakistan* 39, 352-360.

Khatun, A., Hossain, A., Islam, M., Munshi, K., Akter, A., Rahman, B., Huque, R., 2017. Evaluation of gamma irradiation and boiling treatment on antioxidant status in different spices. *J. Food Process Eng.* 40, 1-6.

Kim, J. H., Ahn, H. J., Yook, H. S., Kim, K. S., Rhee, M. S., Ryu, G. H., Byun, M. W., 2004. Color, flavor, and sensory characteristics of gamma-irradiated salted and fermented anchovy sauce. *Radiat. Phys. Chem.* 69, 179-187.

Kim, D. S., Song, M., Kim, S. H., Jang, D. S., Kim, J. B., Ha, B. K., Jeong, I. Y., 2013. The improvement of ginsenoside accumulation in Panax ginseng as a result of γ-irradiation. *J. Ginseng Res.* 37, 332-340.

Kim, M. Y., Kim, G. R., Kim, D. S., Jang, H. K., Kwon, J. H., 2015. Identification characteristics of gamma-irradiated dried fishery and mollusks products using electron spin resonance spectroscopy. *J. Korean Soc. Food Sci Nutr.* 44, 725-731.

Kirkin, C., Mitrevski, B., Gunes, G., Marriott, P. J., 2014. Combined effects of gamma-irradiation and modified atmosphere packaging on quality of some spices. *Food Chem.* 154, 255-261.

Komolprasert, V., 2014. Evaluating packaging materials for use during the irradiation of prepackaged food. In: *Food Additives and Packaging*. American Chemical Society, pp. 119-126.

Komolprasert, V., 2016. Packaging food for radiation processing. *Radiat. Phys. Chem.* 129, 35-38.

Komolprasert, V., Turowski, P. (Eds.), 2014. Food additives and packaging. *American Chemical Society*.

Lee, J. H., Sung, T. H., Lee, K. T., Kim, M. R., 2004. Effect of gamma-irradiation on color, pungency, and volatiles of Korean red pepper powder. *J. Food Sci.* 69, C585-C592.

Lee, H. G., Jo, Y., Kwon, J. H., 2016. Cold pasteurization of frozen crushed garlics using electron beam irradiation and the stability of bioactive components. *Korean J. Food Sci. Tech.* 48, 9-14.

Li, S., Kundu, D., Holley, R. A. 2015. Use of lactic acid with electron beam irradiation for control of *Escherichia coli* O157: H7, non-O157 VTEC *E. coli*, and *Salmonella* serovars on fresh and frozen beef. *Food Microbiol.* 46, 34-39.

Lung, H. M., Cheng, Y. C., Chang, Y. H., Huang, H. W., Yang, B. B., Wang, C. Y. 2015. Microbial decontamination of food by electron beam irradiation. *Trends Food Sci. Technol.* 44(1), 66-78.

Luo, X., Qi, L., Liu, Y., Wang, R., Yang, D., Li, K., Wang, L., Li, Y., Zhang, Y., Chen, Z., 2017. Effects of electron beam irradiation on zearalenone and ochratoxin A in naturally contaminated corn and corn quality parameters. *Toxins* 9, 1-11.

Machi, S., 2004. Emerging applications of radiation technology. In: *TECDOC-1386*. IAEA, Vienna, pp. 5.

Madera-Santana, T. J., Meléndrez, R., González-García, G., Quintana-Owen, P., Pillai, S. D., 2016. Effect of gamma irradiation on physicochemical properties of commercial poly (lactic acid) clamshell for food packaging. *Radiat. Phys. Chem.* 123, 6-13.

Mahalik, N. P., Nambiar, A. N., 2010. Trends in food packaging and manufacturing systems and technology. *Trends Food Sci. Technol.* 21, 117-128.

Mahto, R., Das, M., 2014. Effect of gamma irradiation on the physico-mechanical and chemical properties of potato (*Solanum tuberosum* L.), cv. 'Kufri Sindhuri,' in non-refrigerated storage conditions. *Postharvest Biol. Technol.* 92, 37-45.

Mahto, R., Das, M., 2015. Effect of γ irradiation on the physico-mechanical and chemical properties of potato (*Solanum tuberosum* L), cv. 'Kufri Chandramukhi' and 'Kufri Jyoti,' during storage at 12°C. *Radiat. Phys. Chem.* 107, 12-18.

Mai-Prochnow, A., Murphy, A. B., McLean, K. M., Kong, M. G., Ostrikov, K. K., 2014. Atmospheric pressure plasmas: Infection control and bacterial responses. *Int. J. Antimicrob. Agents* 43, 508-517.

Majchrzak, T., Wojnowski, W., Dymerski, T., Gębicki, J., Namieśnik, J., 2017. Electronic noses in classification and quality control of edible oils: A review. *Food Chem.* 246, 192-201.

Marathe, S. A., Deshpande, R., Khamesra, A., Ibrahim, G., Jamdar, S. N., 2016. Effect of radiation processing on nutritional, functional, sensory and antioxidant properties of red kidney beans. *Radiat. Phys. Chem.* 125, 1-8.

Marcazzó, J., Sanchez-Barrera, C. E., Urbina-Zavala, A., Cruz-Zaragoza, E., 2015. Photostimulated luminescence detection and radiation effects on cinnamon (*Cinnamomum zeylanicum*) spice. *Appl. Radiat. Isotopes* 104, 29-33.

Mohácsi-Farkas, C., Nyirő-Fekete, B., Daood, H., Dalmadi, I., Kiskó, G., 2014. Improving microbiological safety and maintaining sensory and nutritional quality of pre-cut tomato and carrot by gamma irradiation. *Radiat. Phys. Chem.* 99, 79-85.

Morris, C., Brody, A. L., Wicker, L., 2007. Non-thermal food processing/preservation technologies: A review with packaging implications. *Packag. Technol. Sci.* 20, 275-286.

Ndoti-Nembe, A., Vu, K. D., Doucet, N., Lacroix, M., 2013. Effect of combination of essential oils and bacteriocins on the efficacy of gamma radiation against *Salmonella typhimurium* and *Listeria monocytogenes*. *Int. J. Radiat. Biol.* 89, 794-800.

Nielsen, T., Jägerstad, M., 1994. Flavour scalping by food packaging. *Trends Food Sci. Technol.* 5, 353-356.

Pankaj, S. K., Bueno-Ferrer, C., Misra, N. N., Milosavljević, V., O'Donnell, C. P., Bourke, P., Keener, K. M., Cullen, P. J., 2014. Applications of cold plasma technology in food packaging. *Trends Food Sci. Technol.* 35, 5-17.

Park, E. J., Jang, H. N., Jo, D., Kim, G. R., Kwon, J. H., 2013. Physicochemical quality and luminescence characteristics of gamma-irradiated dried fish products. *Korean J. Food Sci. Technol.* 45, 167-173.

Pillai, S. D., Shayanfar, S. (Eds.), 2014. *Electron beam pasteurization and complementary food processing technologies*. MA, USA: Woodhead Publishers, Elsevier.

Pillai, S. D., Shayanfar, S., 2017. Electron beam technology and other irradiation technology applications in the food industry. *Top. Curr. Chem.* 375, 6. https://doi.org/10.1007/s41061-016-0093-4.

Pinela, J., Barreira, J. C., Barros, L., Verde, S. C., Antonio, A. L., Carvalho, A. M., Oliveira, M. B. P., Ferreira, I. C. 2016. Suitability of gamma irradiation for preserving fresh-cut watercress quality during cold storage. *Food Chem. 206*, 50-58.

Polovka, M., Suhaj, M., 2013. Classification and prediction of γ-irradiation of ten commercial herbs and spices by multivariate evaluation of properties of their extracts. *J. Food Nutr. Res.* 52, 45-60.

Praveen, C., Dancho, B. A., Kingsley, D. H., Calci, K. R., Meade, G. K., Mena, K. D., Pillai, S. D., 2013. Susceptibility of murine norovirus and hepatitis A virus to electron beam irradiation in oysters and quantifying the reduction in potential infection risks. *Appl. Environ. Microbiol.* 79, 3796-3801.

Przybytniak, G., Kornacka, E. M., Mirkowski, K., Walo, M., Zimek, Z., 2008. Functionalization of polymer surfaces by radiation-induced grafting. *Nukleonika* 53, 89-95.

Qiu, S., Wang, J., Gao, L., 2014. Discrimination and characterization of strawberry juice based on electronic nose and tongue: Comparison of different juice processing approaches by LDA, PLSR, RF, and SVM. *J. Agric. Food Chem.* 62, 6426-6434.

Rashid, M. H. A., Grout, B. W. W., Continella, A., Mahmud, T. M. M., 2015. Low-dose gamma irradiation following hot water immersion of papaya (*Carica papaya* linn.) fruits provides additional control of postharvest fungal infection to extend shelf life. *Radiat. Phys. Chem.* 110, 77-81.

Roberts D., 1998. Preliminary assessment of the effects of the WTO agreement on sanitary and phyto-sanitary trade regulations. *J. Int. Econ. Law* 1, 377-405.

Robertson, G. L., 2016. *Food packaging: Principles and practice*. Boca Raton FL, USA: CRC press.

Roselló-Soto, E., Barba, F. J., Parniakov, O., Galanakis, C. M., Lebovka, N., Grimi, N., Vorobiev, E., 2015. High voltage electrical discharges, pulsed electric field, and ultrasound assisted extraction of protein and phenolic compounds from olive kernel. *Food Bioprocess Technol.* 8, 885-894.

Sadoughi, N., Karim, R., Hashim, D. M., Zainuri, A., Ghazali, H. M., 2013. Effect of γ-irradiation on the physicochemical properties, and microbial and sensory qualities of cold-stored onion puree. *J. Food Process. Preserv.* 37, 889-898.

Sadoughi, N., Karim, R., Hashim, D. M., Zainuri, A., Ghazali, H. M., 2015. Combined effects of γ-irradiation and ascorbic acid on the physicochemical properties, microbial stability and aroma profile of onion puree during storage. *J. Food Process. Preserv.* 39, 645-652.

Salafranca, J., Clemente, I., Isella, F., Nerín, C., Bosetti, O., 2015. Influence of oxygen and long term storage on the profile of volatile compounds released from polymeric multilayer food contact materials sterilized by gamma irradiation. *Anal. Chim. Acta.* 878, 118-130.

Salvatore, M., Marra, A., Duraccio, D., Shayanfar, S., Pillai, S. D., Cimmino, S., Silvestre, C., 2016. Effect of electron beam irradiation on the properties of polylactic acid/montmorillonite nanocomposites for food packaging applications. *J. Appl. Polym. Sci.* 133. doi: 10.1002/app.42219.

Santos, J. P., Lozano, J., Aleixandre, M., Arroyo, T., Cabellos, J. M., Gil, M., Horrillo, Mdel C., 2010. Threshold detection of aromatic compounds in wine with an electronic nose and a human sensory panel. *Talanta* 80, 1899-1906.

Sanyal, B., Ahn, J. J., Maeng, J. H., Kyung, H. K., Lim, H. K., Sharma, A., Kwon, J. H., 2014. An improved approach to identify irradiated spices using electronic nose, FTIR, and EPR spectroscopy. *J. Food Sci.* 79, C1656-C1664.

Severino, R., Vu, K. D., Donsì, F., Salmieri, S., Ferrari, G., Lacroix, M., 2014a. Antimicrobial effects of different combined non-thermal treatments against *Listeria monocytogenes* in broccoli florets. *J. Food Eng.* 124, 1-10.

Severino, R., Vu, K. D., Donsì, F., Salmieri, S., Ferrari, G., Lacroix, M., 2014b. Antibacterial and physical effects of modified chitosan based-coating containing nanoemulsion of mandarin essential oil and three non-thermal treatments against *Listeria innocua* in green beans. *Int. J. Food Microbiol.* 191, 82-88.

Severino, R., Ferrari, G., Vu, K. D., Donsì, F., Salmieri, S., Lacroix, M., 2015. Antimicrobial effects of modified chitosan based coating containing nanoemulsion of essential oils, modified atmosphere packaging and gamma irradiation against *Escherichia coli* O157: H7 and *Salmonella typhimurium* on green beans. *Food Control* 50, 215-222.

Shahbaz, H. M., Akram, K., Ahn, J. J., Kwon, J. H. 2016. Worldwide status of fresh fruits irradiation and concerns about quality, safety, and consumer acceptance. *Crit. Rev. Food Sci.* *56*(11), 1790-1807.

Shaviklo, A. R., 2015. Development of fish protein powder as an ingredient for food applications: A review. *J. Food Sci. Technol.* 52, 648-661.

Silvestre, C., Cimmino, S. (Eds.), 2013. *Ecosustainable polymer nanomaterials for food packaging: Innovative solutions, characterization needs, safety and environmental issues*. Boca Raton FL, USA: CRC Press.

Stoica, M., Alexe, P., Mihalcea, L., 2014. Atmospheric cold plasma as new strategy for foods processing - An overview. *Innovative Rom. Food Biotechnol.* 15, 1-8.

Supriya, P., Sridhar, K. R., Ganesh, S., 2014. Fungal decontamination and enhancement of shelf life of edible split beans of wild legume *Canavalia maritima* by the electron beam irradiation. *Radiat. Phys. Chem.* 96, 5-11.

Takala, P. N., Vu, K. D., Salmieri, S., Lacroix, M., 2011. Effect of antimicrobial coatings on the radiosensitization of *Escherichia coli, Salmonella typhimurium*, and *Listeria monocytogenes* in fresh broccoli. *J. Food Prot.* 74, 1065-1069.

Tao, Y., Wang, C., Gao, Y., Li, X., Cui, Z., Meng, X. 2017. Effects of different doses of ^{60}Co γ irradiation on storage quality of blueberry fruits. *J. Food Saf. Qual. 8*(7), 2779-2786.

Tawema, P., Han, J., Vu, K. D., Salmieri, S., Lacroix, M., 2016. Antimicrobial effects of combined UV-C or gamma radiation with natural antimicrobial formulations against *Listeria monocytogenes, Escherichia coli* O157: H7, and total yeasts/molds in fresh cut cauliflower. *LWT-Food Sci. Technol.* 65, 451-456.

Tiwari, B. K., O'donnell, C. P., Cullen, P. J., 2009. Effect of non thermal processing technologies on the anthocyanin content of fruit juices. *Trends Food Sci. Technol.* 20, 137-145.

Tsai, S. Y., Mau, J. L., Huang, S. J., 2014. Enhancement of antioxidant properties and increase of content of vitamin D_2 and non-volatile components in fresh button mushroom, *Agaricus bisporus* (higher *Basidiomycetes*) by γ-irradiation. *Int. J. Medicinal Mushrooms* 16, 137-147.

Turgis, M., Millette, M., Salmieri, S., Lacroix, M., 2012. Elimination of *Listeria* inoculated in ready-to-eat carrots by combination of antimicrobial coating and γ-irradiation. *Radiat. Phys. Chem.* 81, 1170-1172.

Uchida, S., 2008. *Utility & advantages of electron beam curing in flexible packaging*. San Jose State University, Packaging Program, San Jose, CA, US.

Valencia-Chamorro, S. A., Pérez-Gago, M. B., Del Río, M. A., Palou, L., 2010. Effect of antifungal hydroxypropyl methylcellulose-lipid edible composite coatings on *Penicillium* decay development and postharvest quality of cold-stored "Ortanique" mandarins. *J. Food Sci.* 75, S418-S426.

Van Bossuyt, M., Van Hoeck, E., Raitano, G., Manganelli, S., Braeken, E., Ates, G., Vanhaecke, T., Van Miert, S., Benfenati, E., Mertens, B., Rogiers, V., 2017. (Q)SAR tools for priority setting: A case study with printed paper and board food contact material substances. *Food Chem. Toxicol.* 102, 109-119.

Vardhan, P. V., Shukla, L. I., 2017. Gamma irradiation of medicinally important plants and the enhancement of secondary metabolite production. *Int. J. Radiat. Biol.* 93, 967-979.

Wesoły, M., Kluk, A., Sznitowska, M., Ciosek, P., Wróblewski, W., 2016. Influence of experimental conditions on electronic tongue results – case of valsartan minitablets dissolution. *Sensors* 16, 1353. doi: 10.3390/s16091353.

WHO, 1999. High-dose irradiation: Wholesomeness of food irradiated with doses above 10 kGy. In: *Report of a Joint FAO/IAEA/WHO Study Group*. Switzerland: World Health Organization.

Wilson, A. D., Baietto, M., 2009. Applications and advances in electronic-nose technologies. *Sensors* 9, 5099-5148.

Yam, K. L., Lee, D. S. (Eds.), 2012. *Emerging food packaging technologies: Principles and practice*. Elsevier.

Yong, H. I., Kim, H. J., Nam, K. C., Kwon, J. H., Jo, C. 2015. Radiation sensitivity of foodborne pathogens in meat byproducts with different packaging. *Radiat. Phys. Chem.* 115, 153-157.

Yun, H., Lacroix, M., Jung, S., Kim, K., Lee, J. W., Jo, C., 2011. Effect of several food ingredients on radiation inactivation of *Escherichia coli* and *Listeria monocytogenes* inoculated into ground pork. *Radiat. Phys. Chem.* 80, 994-997.

Zaman, S., Alam, M. K., Mortuza, M. F., Bari, M. L., 2015. Effectiveness of irradiation treatment in eliminating *E. coli* O157: H7 and *Salmonella* in dried organic herb samples intended for use in blended tea. *J. Food Nutr. Sci.* 3, 165-170.

Zhan, G., Li, B., Gao, M., Liu, B., Wang, Y., Liu, T., Ren, L. 2014. Phytosanitary irradiation of peach fruit moth (*Lepidoptera*: *Carposinidae*) in apple fruits. *Radiat. Phys. Chem. 103*, 153-157.

Zhang, J. Y., Sun, H. J., Song, I. J., Bae, T. W., Kang, H. G., Ko, S. M., Kwon, Y. I., Kim, I. W., Lee, J., Park, S. Y., Lim, P. O., Kim, Y. H., Lee, H. Y., 2014. Plant regeneration of Korean wild ginseng (*Panax ginseng* Meyer) mutant lines induced by γ-irradiation (^{60}Co) of adventitious roots. *J. Ginseng Res.* 38, 220-225.

Zhang, Q. Q., Rui, X., Guo, Y., He, M., Xu, X. L., Dong, M. S., 2017. Combined effect of polyphenol-chitosan coating and irradiation on the microbial and sensory quality of Carp fillets. *J. Food Sci.* 82, 2121-2127.

Zhao, L., Zhang, Y., Pan, Z., Venkitasamy, C., Zhang, L., Xiong, W., Liu, W., 2018. Effect of electron beam irradiation on quality and protein nutrition values of spicy yak jerky. *LWT-Food Sci. Technol.* 87, 1-7.

In: Ionizing Radiation
Editor: Tamar Reeve

ISBN: 978-1-53613-539-8

Chapter 3

GENETIC INSTABILITY OF HAPLOID AND DIPLOID CELLS SURVIVING AFTER EXPOSURE TO UV LIGHT, GAMMA-RAYS AND ALPHA-PARTICLES

Ekaterina S. Evstratova and Vladislav G. Petin*

Biophysical laboratory, Department of Combined Action,
A. Tsyb Medical Radiological Research Centre,
Branch of the National Medical Research Radiological Centre of the
Ministry of Health of the Russian Federation, Obninsk, Russia

ABSTRACT

New results concerning cell survival and genetic instability of wild-type and radiosensitive yeast cells of *Saccharomyces cerevisiae* surviving after irradiation with ^{60}Co γ-rays, ^{239}Pu α-particles and 254 nm UV light are presented. Survival was determined by cell ability to produce

* Corresponding Author Email: vgpetin@yahoo.com.

macrocolonies on a solid nutrient medium. The diploid strain exhibited sigmoid shape of the dose-effect curve independently of radiation type. The haploid strain showed exponential survival curve both after sparsely and densely ionizing radiations while the shape of survival curves was sigmoid after UV exposure. Genetic instability was defined by the delayed appearance of clones by cells surviving irradiation. The effect of delayed colony appearance by irradiated cells was revealed to be more expressed for the diploid wild-type cells characterizing by a sigmoid shape of the dose-effect curve and capable of the diploid-specific recovery in contrast to isogenic haploid strain (100% vs. 20-30%) incapable of such kind of recovery and characterizing by the exponential survival curve. Similar experiments were performed for haploid and homozygous diploid radiosensitive mutants showing the exponential dose-effect curves. On the contrary, UV sensitive mutant exhibits the sigmoid dose-effect curve like a diploid strain. Here again the degree of genetic instability was mainly determined by cell ploidy since both resistant and radiosensitive diploid strains in contrast to haploids demonstrate large extent of the delayed appearance of clones by cells surviving after irradiation with UV light (100% vs. 30%). This means that, in spite of the defects in the repair of DNA double strand breaks, the irradiated cells of the diploid but not haploid strains inherit in a greater extent of some sub-lesions responsible for genetic instability.

Relative biological effectiveness (RBE) of α-particles defined by the ratio of isoeffective dose of γ-rays и α-particles was shown to be of 4.6 ± 0.2 for wild-type diploid cells and only 2.0 ± 0.2 for all haploid strains and radiosensitive diploid strains. It indicates that RBE values are greatly determined by cell ability to recover from radiation damage. These findings are not new for cell survival, while they are fundamentally new for genetic instability.

The dependence of the delayed colony appearance of diploid and haploid yeast cells on their survival was substantially the same after exposure to UV-light, sparsely and densely ionizing radiation. Since at the same survival after exposures to radiations of various quality equally effective number of lethal damage was induced, these data indicate that the identical number of the accompanying sub-lesions responsible for the delayed colony appearance are produced in irradiated cells.

The final conclusion is that genetic instability is mainly determined by cell ploidy rather than the shape of survival curve and the ability of cell to recover from radiation damage as it is traditionally assumed for *Saccharomyces cerevisiae* yeast cells.

Keywords: genetic instability, yeast cells, radiosensitive mutants, ploidy, gamma rays, alpha particles

INTRODUCTION

Genetic instability refer to a set of events capable of causing some alterations within the genome. This term encompasses diverse genetic changes, which can be classified in a variety of ways (Yuen et al., 2007; Zhu et al., 2012; Skoneczna et al., 2015). In particular, a lot of the delayed effects such as chromosomal aberrations, reproductive cell death, genome rearrangement, malignant transformation, reduced cloning efficiency, and heterogeneity among the progeny of irradiated cells, exemplified genetic instability of irradiated cell (Stirling et al., 2011; Kumaran et al., 2013; Lee et al., 2015). Skoneczna et al., (2015) discussed various factors that may influence genome stability, such as cellular ploidy, the phase of the cell cycle, transcriptional activity of a particular region of DNA, the proficiency of DNA quality control systems, the metabolic stage of the cell and its respiratory potential, and finally potential exposure to endogenous or environmental stress.

Chromosome instability appears to be important for the formation of many tumors. Yeast species provide an excellent model for studying ploidy. The yeast *Saccharomyces cerevisiae* has served as an excellent model system for studying pathways involved in genome instability (Duffy et al., 2016). *Saccharomyces cerevisiae* as a model system to delineate the relationship between stress and genetic instability was utilized in paper (Degtyareva et al., 2008). To investigate mechanism by which aneuploidy could contribute to tumorigenesis, the effects of aneuploidy on genomic stability was investigated (Sheltzer et al., 2011). The authors analyzed 13 budding yeast strains that carry extra copies of single chromosomes and found that all aneuploidy strains exhibited one or more forms of genomic instability.

Genetic instability concerns such important phenomena as radiation mutagenesis, carcinogenesis and aging, the main long-term effects of ionizing radiation and UV light (Cheng, Loeb, 1993; Shen, 2011). Genetic instability refers to chromosomal aberrations, reproductive cell death, genome rearrangement, malignant transformation, reduced cloning efficiency, micronucleus formation and heterogeneity among the progeny of irradiated cells. The delayed appearance of clones by cells surviving exposure to ionizing radiation can be also considered as an example of genetic instability of irradiated cell (Chang, Little, 1992; Little, 1998; Dugan et al., 2003; Korogodin et al., 2007). The common property of various delayed effects of radiation is the transmission of sub-lesions to distant progeny. The state of genome instability often precedes tumor progression (Kalb et al., 2006; Shen, 2011) and persists a long time after irradiation over many cell generations (Korogodin et al., 2007; Petin et al., 2008; Klein et al., 1990). A poorly studied aspect of this problem is quantitative comparison of survival and genetic instability for haploid and diploid cells surviving after exposure to carcinogenic factors – γ-rays, α-particles and UV-light.

A role for DNA damage in genetic instability, aging and cancer outcome are under active investigation (Cheng, Loeb, 1993; McMurray, Goltschling, 2004; Shen, 2011). There are evidences that DNA double-strand breaks are important in induction of delayed reproductive death and consequently delayed clone formation by cells surviving exposure to ionizing radiation (Chang, Little, 1992; Brosh, Bohr, 2007; Korogodin et al., 2007). Delayed appearance of colonies along with other delayed effects produced by single cells surviving ionizing radiation exemplifies genetic instability in progeny of irradiated cells. This effect occurs in cultured cells of various origins (Klein et al., 1990; Little, 1998; Morgan, Murnane, 1995). Evidence was presented to support the hypothesis that radiation may induce heritable, genome-wide process of instability that leads to enhanced frequency of genetic change among progeny of original irradiated cell (Klein et al., 1990; Korogodin et al., 2007).

The most convenient object for this purpose is yeast cells (Korogodin, 1966; Kapultcevich et al., 1972). Over the last decade, yeast has been an invaluable model for driving discoveries about the genetic and molecular aspects of aneuploidy. A simple model organism, such as the budding yeast *S. cerevisiae*, has emerged as a robust and versatile model system for studying the effects of genetic alterations. A unicellular eukaryote, the budding yeast is particularly suitable for studying the effects of aneuploidy on cellular physiology because of its small genome, divided into sixteen chromosomes and is well tolerant of aneuploidy (Mulla et al., 2014).

Most results on genetic instability were obtained with cultured mammalian cells in a diploid state. Yeast cells, as the simplest example of eukaryotic cells, are a convenient object for comparing lethal effect and genetic instability of haploid and diploid cells that allows obtaining new fundamental data. In particular, it was shown (Korogodin et al., 2007; Korogodin, 1966; Kapultcevich et al., 1972) that the effect of delayed colony formation by yeast cells surviving after irradiation is more expressed for diploid than for haploid cells. There is evidence that DNA double-strand breaks (DSB) initiate the phenotype of delayed reproductive death (Chang, Little, 1992). It is well known that DNA DSB repair requires two homologous chromosomes (Luchnik et al., 1977; Frankenberg et al., 1981; Karpenhif et al., 2012) and due to this ability diploid yeast cells are more resistant to ionizing radiation than haploids whose ability to repair DSB is greatly reduced and may be realized through nonhomologous end-joining (Lewis, Resnick, 2000; Mao et al., 2008). This implies that radioresistant diploid yeast cells showing the sigmoid dose-effect curves inherit a number of non-lethal sub-lesions responsible for genetic instability. Similar effect was lesser observed for wild-type haploid yeast cells showing the exponential dose-effect curves. On this basis, it was concluded that the shape of survival curve and cell ability to DNA DSB repair determine the delayed appearance of clones. However, in our recent work with haploid and diploid yeast cells of wild-type and their radiosensitive mutants it was noticed (Evstratova et al., 2016) that genetic instability defined by the delayed appearance of clones was well expressed and reached 100% for diploid cells irrespectively of the shape of survival curve and cell ability to recover from radiation damage. This effect was poorly expressed (20–25%) for all haploid strains characterized by an exponential survival curve. On this basis, a preliminary assumption was drawn that the genetic instability induced by ionizing radiation is largely determined by cell ploidy, rather than by the shape of survival curve and cell ability to recover from radiation damage. This assumption needs additional confirmation. It has long been known that the shape of survival curves after irradiation with UV light, unlike the effect of ionizing radiation, does not depend on ploidy and has a sigmoid form for both haploid and diploid yeast cells (Korogodin, 1966). Then a comparison of genetic instability of haploid and diploid yeast cells surviving after UV light exposure could contribute to a better understanding whether this effect is actually related with the shape of survival curve or to a greater extent is determined by cell ploidy.

The role of genetic instability in cellular radioresistance was widely discussed (Morgan, Murnane, 1995; Kaplan, Morgan, 1998; Petin et al., 2008). However, there are some contradictions in this issue. The observation that double-strand breaks (DSB) repair-deficient CHO mutant *xrs*-5 cells do not show delayed reproductive death (delayed appearance of colonies) supports the idea that DSB repair must play an important role in genetic instability (Chang, Little, 1992). Human radiosensitive mutant cell lines such as xeroderma pigmentosum, Fanconi's anemia and ataxia-telangiectasia have DNA repair defects and demonstrate genomic instability (Gatti et al., 1991; Kalb et al., 2006; Sugasawa, 2008). Several authors have investigated genome instability using *Saccharomyces cerevisiae*, a simple single-celled organism model (McMurray, Goltschling, 2004; Sheltzer et al., 2011; Yuen et al., 2007). Delayed appearance of colonies by cells surviving ionizing radiation exposure was more pronounced for diploid than for haploid yeast cells of wild-type (Korogodin, et al., 2007). This means that radioresistant diploid yeast cells showing sigmoidal survival curves inherit a number of non-lethal sub-lesions responsible for genetic instability. Similar effect was small for wild-type haploid yeast cells showing exponential survival curves.

Densely ionizing radiation with high linear energy transfer (LET) seems promising as cancer treatment because of its higher relative biological effectiveness (RBE) in low dose region usually used at fractionated irradiation of malignant tumors and due to the suppression of cell recovery from sublethal and potentially lethal damage (Petin, Kabakova, 1981; Frankenberg et al., 1981; Frankenberg-Schwager et al., 1984; Hall, Giaccia, 2011; Petin, Kapultcevich, 2014). Although the genetic instability is well known after exposure to high LET radiation

(Kadhim et al., 1994; Little et al., 1997; Ponnaiya et al., 2004), RBE of densely ionizing radiation has never been quantified for this effect. The authors of the paper (Dugan, Bedford, 2003) indicates that the frequency of unstable clones was greater after exposure of cultured normal human cells to densely ionizing ^{56}Fe particle nuclei than to ^{137}Cs γ-rays. However, the authors also didn't estimate quantitatively the RBE values. Moreover, nobody tried to compare the RBE of densely ionizing particles for both cell inactivation and genetic instability. However, it is not excluded that the effect of the delayed appearance of colonies may be not related to survival curve shape and completely is determined by cell ploidy. There were no publications related to comparing of genetic instability for haploid and diploid wild-type yeast cells and their radiosensitive mutants. Therefore, it would be of interest to compare quantitatively the delayed appearance of clones produced by haploid and diploid wild-type yeast cells and by their radiosensitive mutants surviving radiation exposure. In addition, the comparison of RBE for both lethal effects and genetic instability of haploid and homozygous diploid cells differing in their radiosensitivity remains completely unexplored.

Thus, the following purposes should be performed in this work: (i) to determine whether genetic instability is determined by cell ability to recover from radiation damage and the shape of the survival curve, or to a greater extent depends on cell ploidy; (ii) to quantify the RBE of α-particles for cell survival and genetic instability for haploid and diploid yeast cells; (iii) to compare the delayed appearance of clones by cells surviving after exposure to γ-rays, α-particles, and UV-light.

Material and Methods

Strains and Cultivation Condition

The following haploid and homozygous diploid strains of *Saccharomyces cerevisiae* yeasts were used for these experiments: S288C (RAD) and XS800 (RAD/RAD); XS774-4d (*rad*51) and XS806 (*rad*51/*rad*51); g160/2d (*rad*52) and XS1898 (*rad*52/*rad*52); g218/7c (*rad*54) and XS1879 (*rad*54/*rad*54). We also used UV light sensitive diploid yeasts (strain XS1924, *rad*18/*rad*18). The strains with a single mutation were the most radiosensitive among those known at present (Saeki, Machida et al., 1980; Petin, Kim, 2016). Haploid strains of wild-type and their radiosensitive mutants were initially obtained from Dr. R. Mortimer (USA); all diploid strains have been obtained from Dr. S. Nakai (Japan).

Yeast cells has a number of properties that make it possible to study the mechanisms of various processes at the molecular, subcellular, cellular and population levels without using complex technologies and to obtain results in a relatively short time (Korogodin, 1966; Korogodin et al., 2007). The short life cycle of yeast and the possibility of rapidly obtaining a large number of cell generation permits to study even very rare phenomena. Most radiobiological responses of yeast cells, such as survival curve shape, dependence of RBE on LET, oxygen effect, action of radioprotectors and radiosensitizers, are qualitatively similar to those of mammalian cells (Petin et al., 2012; Petin, Kapuitcevich, 2014; Petin, Kim, 2016). Convenience of yeast cell cultivation provides a clear reproducibility of results obtained. In addition, they have other positive qualities: rapid growth on a dense nutrient medium with a large yield of cells; the ability to form homogeneous suspensions consisting of single cells; a small number of budding cells in a population that has reached a stationary stage of growth. The cells were grown before irradiation up to stationary phase on a solid nutrient medium, and then they were washed off with distilled water and resuspended to make a stock solution. The length of pre-irradiation growth was determined by cell budding cease.

Irradiation with Ionizing Radiations and UV Light

Cells from the same suspension were exposed to ^{60}Co γ-rays (LET = 0.2 keV/μm, 20 Gy/min) and ^{239}Pu α-particles (25 Gy/min). The γ-ray dose rate was measured with a calibrated Siemens ionization chamber. The LET of α-particles reaching a cell monolayer was estimated to be of 120–130 keV/μm. Exactly at about this LET value the maximum of RBE-LET relationship was observed for most eukaryotic and some prokaryotic unicellular organisms (Raju, 1980; Hall, Giaccia, 2011). The α-particle dose rate was determined by measuring of the intensity and energy of the particles with a semi-conductor silicon surface barrier detector at a distance corresponding to the cells. Procedure for α-irradiation was similar to that described earlier (Petin, Kim, 2005; Petin, Kapuitcevich, 2014). The small range of α-particles necessitated the use of a monolayer of the yeast cells for α-irradiation: 0.02 ml of cell suspension was placed on the surface of a non-nutrient agar, and the water from this drop of suspension was evaporated. All irradiations were carried out at room temperature (20 ± 2°C).

The cell suspension (2 ml, 10^6 cells/ml) was irradiated with UV light (254 nm, 1.5 W/m^2) in an open quartz vessel. The fluence rate was measured using a calibrated General Electrical germicidal meter. To avoid photoreactivation, irradiation, dilution of suspensions and other procedures were performed at red light, while post-radiation incubation of cells occurred in dark conditions.

Survival and Genetic Instability Assays

After the treatment, the samples were diluted to the appropriate cell concentration and a known number of cells were plated in such a manner that survival cells would produce 50–300 colonies. Dose-response curves were independent of whether or not the cell suspension was prepared with 0.07 M phosphate buffer or with distilled water. Survival response was determined on the basis of the colony counts obtained at the end of 2–3 days of incubation at 30°C and the counts checked again after a further period to ensure that the final score had been reached. Each data point represents average survival from three to six Petri dishes. Points in all the figures are the means from at least three independent experiments. Error bars in all the figures indicate the standard errors of the mean.

To quantify the delayed formation of colonies by cells surviving after irradiation, the colony grown in Petri dishes was counted after 22 hours and every 2–6 hours later until the colonies ceased to appear. The last count was made after 4 days. As a test of genetic instability, the percentage of colonies formed later than in control was used. The correspondence of such test to genetic instability was confirmed earlier (Korogodin et al., 2007; Petin et al., 2008) by a number of factors – cells form colonies produced later than in control were characterized by increased radio- and thermosensitivity, increased content of morphologically altered colonies, increased content of nonviable cells inside late-formed colonies, and an increased content of respiratory and recombinant mutants. Each experiment was repeated 2–5 times. The results are presented as an average and its standard error.

RESULTS

Figure 1 shows the dependence of cell survival on exposure dose (further for brevity simply "survival curves") for haploid (closed circles) and diploid (open circles) yeast cells exposed to 254 nm UV light (panel A), ^{60}Co γ-rays (panel B), ^{239}Pu α-particles (panel C).

It is clear that irradiation with UV-light (Figure 1, A) haploid and diploid yeast cells exhibited the sigmoid shape of survival curves regardless of cell ploidy. Haploid yeast cells were more sensitive to UV light than the diploid (1.6 times at 10% survival).

After irradiation with sparsely ionizing radiation (Figure 1, B), haploid cells exhibit exponential survival curve while diploid cells showed sigmoid inactivation kinetics and they were more resistant in comparison with haploids (6.2 times at 10% survival). This difference is usually interpreted as being due to repair of DNA double-strand breaks that requires two homologous DNA duplexes (Luchnik et al., 1977; Frankenberg et al., 1981; Frankenberg-Schwager et al., 1984). The exponential shape of survival curve for haploid cells indicates a less efficiency of recovery from radiation-induced damage. The ability of haploid cells to recover double-strand breaks is also known by virtue of discovery of repair by nonhomologous end-joining (Lewis, Resnik, 2000; Mao et al., 2008), the effectiveness of which is much lower than the recovery through the recombination of homologous chromosomes.

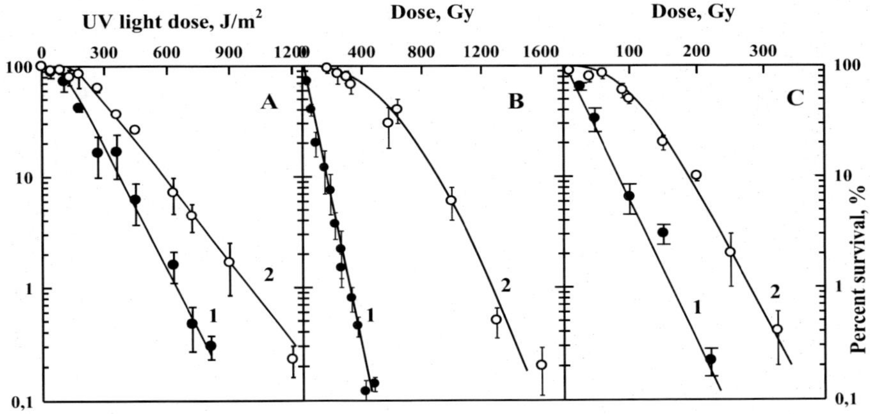

Figure 1. Survival curves of wild-type haploid (curves 1) and diploid (curves 2) *S. cerevisiae* yeast cells exposed to 254 nm UV light (panel A), ^{60}Co γ-rays (panel B), ^{239}Pu α-particles (panel C).

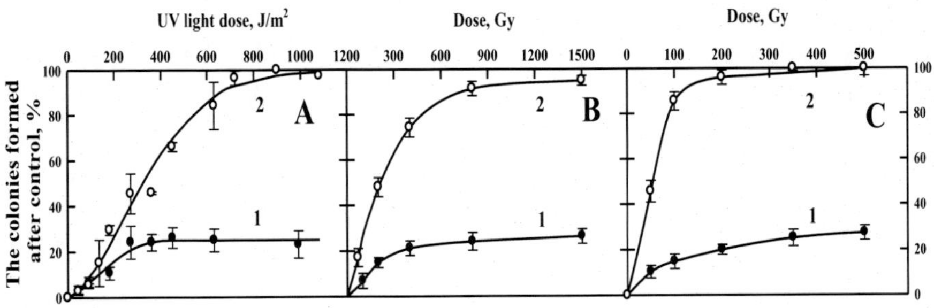

Figure 2. Delayed appearance of colonies produced by haploid (curves 1) and diploid (curves 2) *S. cerevisiae* yeast cells surviving after irradiation with 254 nm UV light (panel A), ^{60}Co γ-rays (panel B) and ^{239}Pu α-particles (panel C).

After exposure to α-particles (Figure 1, C), the shape of survival curves of haploid and diploid cells stayed without change in comparison with gamma irradiation and diploid cells remained more resistant than the haploid, although the degree of curvature of the initial part of diploid survival curve was slightly decreased. The difference in their radioresistance also decreased (2.1 times at 10% survival). This was due to the fact that high-LET radiations produce a greater proportion of irreversible damage, from which cells are unable to recover (Petin, Kabakova, 1981; Blöcher, 1988). In these papers, it was shown that the irreversible component was enhanced for the densely ionizing radiation in comparison to the low-LET radiation while the probability of the recovery was identical for both the low- and high-LET radiations for a lot of yeast strains investigated (Petin, Kim, 2005). It was concluded that the recovery process itself is not damaged after densely ionizing radiation and the enhanced RBE of the high-LET radiation may be caused by the increased yield of the irreversible damage.

The RBE of α-particles for cell survival is defined as the ratio of the mean lethal doses D_0 after exposure to low- and high-LET radiations (RBE = $D_{0\gamma}/D_{0\alpha}$), D_0 being the dose needed to reduce the cell survival to 37% after low- and high-LET radiations. In contrast with haploid cells, survival curves for wild-type diploid cells were sigmoid after exposure both to γ-rays or α-particles. For such shape of survival curves, the RBE was estimated by the ratio of the corresponding D_0 on the exponential parts of the respective survival curves. The RBE values calculated in this way on the basis of the data presented in Figure 1,B and 1,C, were 2.0 for haploid and 4.6 for diploid yeast cells. The greatest value of RBE (4.6) was observed for wild-type diploid cells. Many other diploid yeast cells of wild-type capable to recover from radiation damage revealed similar values of RBE (Frankenberg et al., 1981; Petin, Kabakova, 1981; Blöcher, 1988). In contrast to these data, RBE for haploid yeast cells of wild-type and all mutant strains independently of their ploidy varies accidentally within the range of 1.8–2.7 without any systematic regularity.

As noted earlier, genetic instability, determined by the delayed appearance of colonies by irradiated cells, is more pronounced for diploid cells, regardless of the shape of the survival curve and the ability of cells to recover (Evstratova et al., 2016). In order to confirm this conclusion, it was of interest to compare genetic instability induced by various kinds of radiations in haploid and diploid cells. Figure 2 shows the delayed appearance of clones by haploid (curves 1) and diploid (curves 2) yeasts surviving after exposure to UV light (panel A), ^{60}Co γ-rays (panel B) and ^{239}Pu α-particles (panel C).

It is evident that as the dose of the influencing factor increases, the proportion of surviving diploid cells forming colony later than in control increases and reaches almost 100% only for diploid cells, while this effect was about 30% for haploid cells irrespective of kind of radiation. These data indicates that as the dose increases, the proportion of cells containing ineffective sub-lesions, responsible for the delayed appearance of clones, also increases. It can be seen that the genetic instability of haploid yeast cells also characterized by a sigmoid survival curve for UV irradiation, is much less pronounced than in diploid cells, and is about of 30%. This confirms our earlier conclusion (Evstratova et al., 2016) that genetic instability is mainly determined by cell ploidy rather than by the shape of survival curves.

It is also seen that the delayed clone appearance increased more sharply with dose of α-irradiation than γ-rays that indicates a greater efficiency of manifestation of the detected effect after the action of densely ionizing radiation. The RBE of α-particles, determined by the ratio of the isoeffective doses of γ- and α-radiations, was 4.7 ± 0.2 in the range of the measured parameter changed from 20 to 60%, i.e. α-particles are 4.7 times more effective than the action of sparsely ionizing radiation that corresponds to the RBE of α-particles, estimated by the criterion of cell survival in this paper and reported in other works (Frankenberg-Schwagen et al., 1984; Petin, Kim, 2005; Petin, Kapultcevich, 2014). The same increase in the efficiency of α-particles for cell inactivation and the delayed appearance of clones by surviving cells may be associated with the greater efficiency of densely ionizing radiation to produce lethal radiation damage and the corresponding sub-lesions that persists in distant descendants of surviving cells, thereby destabilizing the genome, which can manifest themselves in the slowing down of their reproduction (Korogodin et al., 2007; Petin et al., 2008).

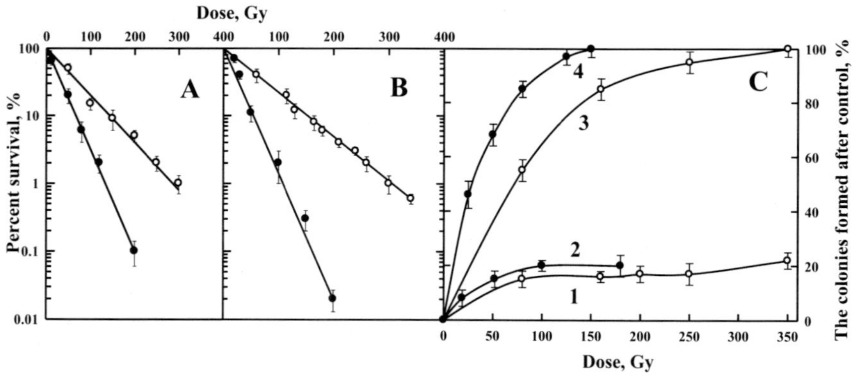

Figure 3. Radiobiological responses of radiosensitive mutants of haploid (strain XS774-4d, *rad*51) and diploid (strain XS806, *rad*51/*rad*51) *S. cerevisiae* yeast cells. A: survival-versus-dose curves of radiosensitive haploid yeasts exposed to ^{60}Co γ-rays (open circles) and ^{239}Pu α-particles (closed circles). B: survival-versus-dose curves of radiosensitive diploid yeasts exposed to ^{60}Co γ-rays (open circles) and ^{239}Pu α-particles (closed circles). C: the delayed appearance of colonies produced by radiosensitive haploid (curves 1, 2) and diploid (curves 3, 4) yeast cells surviving after irradiation with ^{60}Co γ-rays (open circles) and ^{239}Pu α-particles (closed circles).

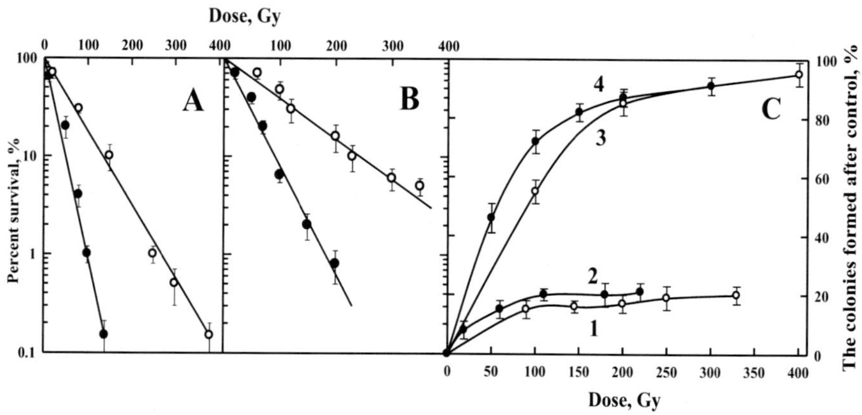

Figure 4. The same as in Figure 3, but for radiosensitive mutants of haploid (strain g160/2d, *rad*52) and diploid (strain XS1898, *rad*52/*rad*52) *S. cerevisiae* yeast cells.

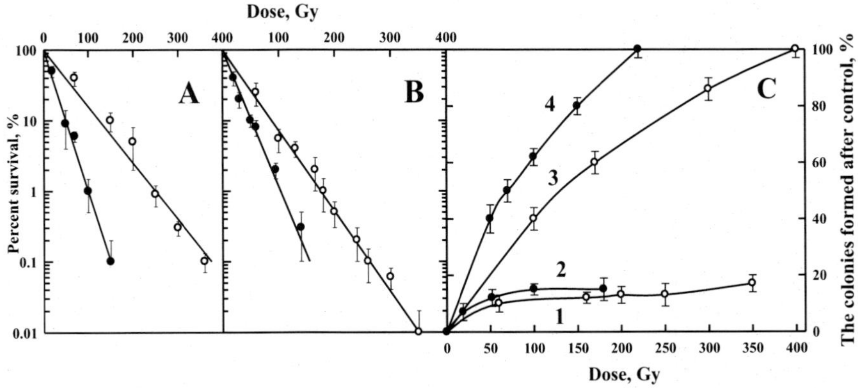

Figure 5. The same as in Figure 3, but for radiosensitive mutants of haploid (strain g218/7c, *rad*54) and diploid (strain XS1879, *rad*54/*rad*54) *S. cerevisiae* yeast cells.

Figure 3 shows survival curves for haploid (A) and diploid (B) mutants yeast cells (RAD51 loci) irradiated with γ-rays (open circles) and α-particles (closed circles). After both kinds of irradiation, both haploid and diploid cells exhibit exponential survival curve. Analogous results obtained with other strains are presented in Figure 4 and Figure 5 (RAD52 and RAD54 loci respectively). These mutants lost their ability to recover from radiation damage mainly due to molecular defects in recombination process and were inactivated exponentially independently of their ploidy after exposure to low- and high-LET radiation.

Figures 3–5 (panels C) exhibit delayed appearance of clones by haploid (curves 1, 2) and diploid (curves 3, 4) yeast cells surviving after exposure to γ-rays (open circles) and α-particles (closed circles). The delayed appearance of colonies was quantitatively compared with unirradiated control cells. It is clear from these data that exposure to high-LET radiation was more effective for genetic instability for all yeast strains studied as it was reported by other investigators for cultured mammalian cells (Kadhim et al., 1994; Little et al., 1997; Ponnaiya, 2004). However, our findings, in contrast to these papers, allow quantifying the RBE values, which were slightly dependent on exposure dose. By averaging these results, we get RBE values for genetic instability. Here again, the greatest value of RBE (4.7) was observed for wild-type diploid cells while for haploid yeast cells of wild-type and all mutant strains independently of their ploidy RBE varies accidentally within the range of 1.8–2.7 without any systematic regularity. It is also evident that delayed appearance of clones, as a measure of radiation-induced genetic instability, is dose-dependent effect and more pronounced for all diploid cells (up to 100%) independently of their radiosensitivity then for haploid cells (up to 20-30%). This pattern is common for all diploid and haploid strains independently of the shape of dose-response curves and cell ability to recover from radiation-induced damage.

We performed an experimental comparison of wild-type and UV sensitive strains. It turned out that the genetic instability of the UV sensitive strain is much larger than that of the wild-type cells (Figure 6).

In order to compare the effectiveness of ionizing radiation action on genetic instability with the influence of UV light on this effect, we presented the dependence of the delayed colony formation on cell survival for haploid and diploid yeast cells (Figure 7). This figure contains data for the late appearance of colonies by cells surviving after irradiation with γ-rays, α-particles, and UV-light. The obtained results show that the delay in the formation of colonies by irradiated cells in the dependence on their survival was practically independent of kind of radiation. These data indicate that equally effective lethal damage is accompanying by equal number of sub-lesions responsible for genetic instability.

In other words, this effect is associated with the accumulation of sub-lesions ineffective for cell inactivation. We can note again these sub-lesions are preserved in the remote descendants of surviving cells after irradiation and may manifest themselves in slowing down of their reproduction (Korogodin et al., 2007; Petin et al., 2008). This problem is important in connection with the use of synergy ideas (Petin, Kim, 2016) in the treatment of tumors in photoradiation and photochemotherapy (Melloni et al., 1984).

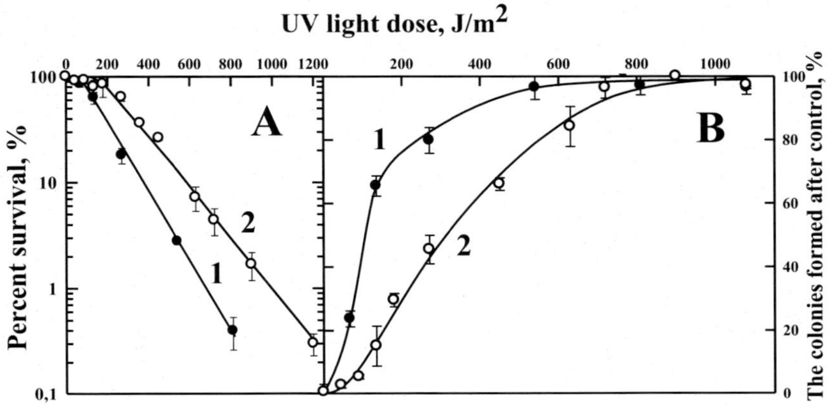

Figure 6. Dependence of the survival (A) and the delayed appearance of colonies (B) for UV sensitive diploid (strain XS1924, *rad*18/*rad*18, curves 1) and wild-type (strain XS800, curves 2) of *Saccharomyces cerevisiae* yeast cells on the dose of UV light.

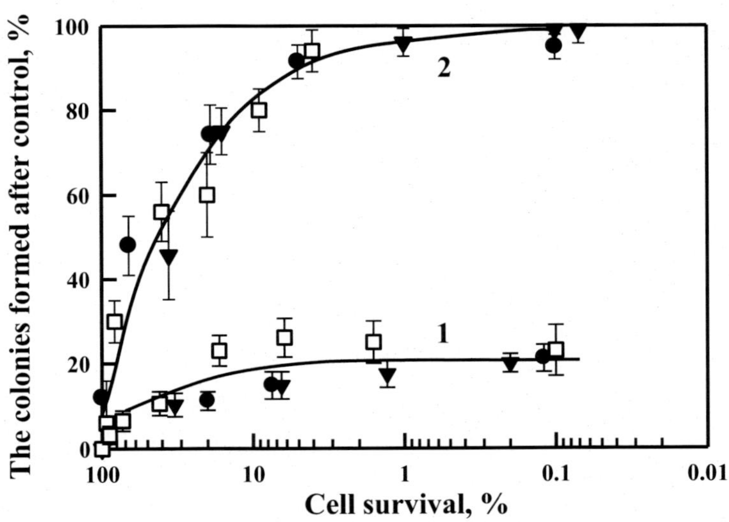

Figure 7. The relationship between cell survival and the delayed appearance of colonies produced by haploid (curve 1) and diploid (curve 2) *S. cerevisiae* yeast cells surviving after irradiation with ^{60}Co γ-rays (circles), ^{239}Pu α-particles (triangles) and 254 nm UV light (squares).

DISCUSSION

The delayed appearance of clones by cells surviving exposure to ionizing radiation, like other delayed effects such as chromosomal aberrations, reproductive cell death, genome rearrangement, malignant transformation, reduced cloning efficiency, micronucleus formation and heterogeneity among the progeny of irradiated cells, exemplified genetic instability of irradiated cell (Morgan, Murnane, 1995; Yuen et al., 2007; Shen, 2011; Aggarwal, Brosh, 2012; Skoneczna et al., 2015). The results obtained for survival and delayed appearance of clones by isogenic haploid and diploid *Saccharomyces cerevisiae* yeast cells of wild-type surviving after exposure to various radiations exemplifying genetic instability are partly confirmed by previously published data (Korogodin, 1966; Kapultcevich et al., 1972; Korogodin et al., 2007; Petin et al., 2008) indicating this effect is more pronounced for diploid than for haploid yeast strains (100% vs. 20%). It was asserted in the cited papers that genetic instability is natural for diploid yeast cells with sigmoid shape of dose-response curves due to cell ability to recover from radiation damage, while this effect was negligible or even not observed for strains with exponential survival curves. It might be the time to shift this paradigm. Indeed, the authors of the cited paper didn't use radiosensitive mutants with exponential survival curves both for haploid and diploid yeast cells exposed to low- and high-LET radiations. We suggested that genetic instability may be essentially determined by yeast cell ploidy and was not related with the shape of dose-response curve and cell ability to recover from radiation-induced damage. To test this suggestion, we compared the delayed appearance of clones produced by the haploid and homozygous diploid wild-type and radiosensitive mutants surviving exposure to radiations with different LET. It was shown that both resistant and radiosensitive diploid strains in contrast to haploids demonstrate large extent of genetic instability (100% vs. 20-30%) independently of radiation quality.

In this relation, it is interesting that recent studies in cancer cells and budding yeast demonstrated that aneuploidy, the state of having abnormal chromosome numbers, correlates with elevated chromosome instability, i.e., the propensity of gaining and losing chromosomes at a high frequency. correlates with cellular ploidy in an unexpected way: cells with a chromosomal content close to the haploid state are significantly more stable than cells displaying an apparent ploidy (Zhu et al., 2012). Accumulating evidence suggests that an increased number of chromosome sets increases genomic instability while improving cellular capacity to buffer the detrimental effects of gene mutations or chromosome loss. The evidence presented supports the idea that increased ploidy accelerates chromosome missegregation, with a strong bias to chromosome loss, and promotes chromosomal rearrangements, while the extra sets of chromosomes allow survival of the resulting aneuploid cells while buffering the effects of detrimental chromosome loss or gene mutation. These unique features can generate cells with variable karyotypes, thus providing material for natural selection to act upon (Lee et al, 2015).

Despite the defects, blocking repair of double-strand breaks (DSB) in radiosensitive diploid strains, mutant cells surviving irradiation inherit residual sub-lesions responsible for the manifestation of genetic instability in the same extent as the wild-type diploid yeast cells capable of DSB recovery. This implies the genetic instability caused by ionizing radiation, is largely determined by cell ploidy, rather than the shape of the dose-effect curve and cell ability to recover from radiation damage. This point of view is supported by the data presented here on the genetic instability of yeast cells after UV irradiation. Despite the sigmoid form of survival curves for both haploid and diploid yeasts, the delayed appearance of clones is more pronounced for diploid but not for haploid cells (100% vs. 30%). Consequently, the effect of the delayed appearance of clones by diploid yeast cells surviving after irradiated is not uniquely related to the shape of survival curve and cell ability to recover from radiation damage. In other words, the presence of the diploid-specific repair (Saeki et al., 1980) is not the main determinant of the late appearance of colonies by irradiated cells.

The dependence of the delayed colony appearance by haploid and diploid yeast cells on their survival was substantially the same after exposure to UV-light, sparsely and densely ionizing radiation. On this basis, a nontrivial conclusion is proposed consisting in that the number of the produced radiation damage responsible for cell death may be directly proportional to the number of some sub-lesions accountable for genetic instability.

Thus, the data presented here and some of our earlier data (Evstratova et al., 2016) show that the delayed appearance of clones produced by cells surviving after irradiation with ionizing radiations of different quality and UV light is mainly determined by cell ploidy rather than by the sigmoid form of survival curve and cell ability to recover radiation damage, as repeatedly postulated (Korogodin et al., 2007; Korogodin, 1966; Kapultcevich et al., 1972). Yeast cells are the simplest model of eukaryotic cells. If we assume that these patterns can be retained for the higher eukaryotes, one would expect that the genetic instability would be more expressed for diploid somatic cells than for haploid sex cells.

Another important new fact obtained here is that RBE for cell survival of wild-type diploid yeast cells capable to repair radiation-induced damage is 4.6 ± 0.2 and not significantly different from RBE (4.7 ± 0.4) for the delayed appearance of colonies. RBE for cells with greatly reduced ability to recover from radiation damage was significantly smaller for both survival and genetic instability. The values of RBE varied accidentally within the range of 1.8–2.7 without any systematic regularity for haploid wild-type yeast cells and all mutant cells independently of their ploidy. These findings are not new for cell survival, while they are fundamentally new for genetic instability. Identity of RBE values for cell survival and genetic instability may be associated with the greater efficiency of densely ionizing radiation to produce lethal radiation damage and accompanying sub-lesions responsible for the delayed appearance of clones by cells surviving after irradiation. Sub-lesions produced are preserved in the remote descendants of cells surviving after irradiation and thereby cause destabilization of genome, which is manifested in the delayed appearance of clones after irradiation. These RBE values for cell survival have been published by many authors (Petin, Kabakova, 1981; Frankenberg et al., 1981; Frankenberg-Schwager et al., 1984; Petin, Kim, 2005; Petin,

Kapultcevich, 2014) while the RBE magnitude for genetic instability are presented for the first time. The explanation of these results can be originated from the fact that high-LET radiations are known to produce clustering of ionizations and excitations in DNA (Goodhead, 1994), which produce more severe and complex damage than low-LET reference radiation. This also agrees with our earlier results (Petin, Kim, 2005) that densely ionizing radiation inhibits cell recovery due to greater fraction of irreversible damage, while the recovery constant did not depend on radiation quality.

As a whole, the data presented here indicate that RBE of densely ionizing radiation is substantially determined by the ability of cells to recover from radiation damage both for cell survival and genetic instability. The unexpected role of the specific repair pathways in the RBE values of high-LET radiation was reaffirmed here for cell survival and was demonstrated for genetic instability for the first time. The influence of densely ionizing radiation on cellular survival observed in this paper confirm the well-known data that RBE of densely ionizing radiation is determined by both physical properties of radiation and yeast cell capability to recover radiation damage (Frankenberg et al., 1981; Petin, Kabakova, 1981; Blöcher, 1988). Such a recovery may be mediated by a recombination-like mechanism, for which the diploid state is required (Frankenberg et al., 1981; Petin, Kabakova, 1981). Similar explanation may be engaged for RBE regarding the delayed appearance of clones.

Although biological effects associated with radiation-induced genetic instability are widely discussed, the natures of events that initiate and retain instability in the progeny of irradiated cells as yet remain unclear. Careful discussion of molecular mechanisms underlying genetic instability described here are beyond the main scope of this work. However, some points may be mentioned. Many mechanisms have been proposed on how radiation exposure can give rise to genetic instability (Aggarwal, Brosh, 2012; Skoneczna et al., 2015).

While it was shown the nucleus is the target for radiation-induced genetic instability, it is likely that single gene, gene family, a single radiation-induced mutation or DNA double-strand breaks are not sufficient for the induction genetic instability (Kaplan, Morgan, 1998). Eukaryotic cells possess two DNA double-strand break pathways: homologous recombination (Luchnik et al., 1977; Frankenberg et al., 1981; Karpenshif, Bernstein, 2012) and nonhomologous DNA end joining (Lewis, Resnick, 2000; Mao et al., 2008). Unrepaired or misrepaired DNA double-strand break may result in both cell inactivation and genetic instability. It was supposed that the repair of radiation induced DNA double strand breaks may initiate the genomic instability and carcinogenesis (Klein et al., 1990; Chang, Little, 1992; Cheng, Loeb, 1993). It was noted that mitotic recombination can serve as one of the ways of genome instability initiation (Tolstorukov et al., 1979). Defects in the DNA repair process are known to lead to DNA damage, which in turn can cause cell cycle arrest, apoptosis, and tumorigenesis. Homologous recombination may be a crucial metabolic pathway found in all organisms, and it can be involved in genetic instability (Lee et al, 2015). The increased frequency of mitotic recombination may correspond to cells that received upon irradiation some sub-lesions that persist in distant descendants of cells surviving after irradiation and responsible for the delayed appearance of clones. Rad51 and Rad52 are key proteins involved in DSB repair by HR and are conserved between humans and yeast (Lee et al, 2015). Since in our paper the genetic instability was demonstrated for yeast strains defective in recombination (*rad*51/*rad*51, *rad*52/*rad*52), it may be reasonably assumed that homologous recombination cannot be considered as a reason for the delayed appearance of clones observed here. Mechanism of different manifestation of the genetic instability by haploid and diploid yeast cells surviving radiation exposure is independent on cell ability to recover from radiation damage and could be conceivably related with some chromosomal damages, which are lethal for haploid but not for diploid cells. The nature of such kind of damage needs to be elucidated further for more clarification.

It may be concluded that the mechanism of different manifestation of the genetic instability by haploid and diploid yeast cells surviving radiation exposure is independent on cell ability to recover from radiation damage and could be conceivably related with some chromosomal damages, which are lethal for haploid but not for diploid cells. The nature of such kind of damage stays to be elucidated.

REFERENCES

Aggarwal, M; Brosh, RM. Functional analysis of human DNA repair proteins important for aging and genomic stability using yeast genetics. *DNA repair (Amst)*, 2012, 11, 335–348.

Blöcher, D. DNA double-strand break repair determines the RBE of α-particles. *International Journal of Radiation Biology,* 1988, 54, 761–771.

Brosh, RM; Bohr, VA. Human premature aging, DNA repair and RecQ helicases. *Nucleic Acids Research,* 2007, 35, 7527–7544.

Chang, WP; Little, JB. Evidence that DNA double-strand break initiate the phenotype of delayed reproductive death in Chinese hamster ovary cells. *Radiat Res*, 1992, 131, 53–59.

Cheng, KC; Loeb, LA. Genomic instability and tumor progression: mechanistic considerations. *Adv Cancer Res*, 1993, 60, 121–156.

Degtyareva, NP; Chen, L; Mieczkowski, P; Petes, TD; Doetsch, PW. Chronic Oxidative DNA Damage Due to DNA Repair Defects Causes Chromosomal Instability in *Saccharomyces cerevisiae. Molecular and Cellular Biology*, 2008, 28, 17, 5432–5445.

Duffy, S; Fam, HK; Wang, YK; Styles, EB; Kim, J-H; Ang, JS; Singh, T; Larionov, V; Shah, SP; Andrews, B; Boerkoeland, CF; Hieter P. Overexpression screens identify conserved dosage chromosome instability genes in yeast and human cancer. *PNAS* 2016, 113, 36, 9967–9976.

Dugan, LC; Bedford, JS. Are chromosomal instabilities induced by exposure of cultured normal human cells to low- and high-LET radiation? *Radiation Research,* 2003, 159, 301–311.

Evstratova, ES; Pereklad, OV; Petin, VG. The dependence of radiation-induced genetic instability on yeast cell ploidy. *Radiation and Risk*, 2016, 25, 80–89.

Frankenberg, D; Frankenberg-Schwager, M; Blöcher, D; Harbich, R. Evidence for DNA double-strand breaks as the critical lesions in yeast cells irradiated with sparsely and densely ionizing radiation under oxic or anoxic conditions. *Radiat Res*, 1981, 88, 524–532.

Frankenberg-Schwager, M; Frankenberg, D; Harbich, R. Repair of DNA double-strand breaks as a determinant of RBE of alpha-particles. *Br J Cancer,* 1984, 49, 169–73.

Gatti, RA; Boder, E; Vinters, HV; Sparkes, RS; Norman, A; Lange, K. Ataxia-telangiectasia: An interdisciplinary approach to pathogenesis. *Medicine,* 1991, 70, 99–117.

Goodhead, DT. Initial events in the cellular effects of ionizing radiations: clustering damage in DNA. *Int J Radiat Biol*, 1994, 65, 7–17.

Hall, EJ; Giaccia, AJ. *Radiobiology for the Radiologist*. Seventh edition. Philadelphia: Williams and Wilkins, Lippincott; 2011.

Kadhim, MA; Lorimore, SA; Hepburn, MD; Goodhead, DT; Buckle, VJ; Wright, EG. Alpha-particle-induced chromosomal instability in human bone marrow cells. *Lancet*, 1994, 344, 987–988.

Kalb, R; Neveling, K; Nanda, I; Schindler, D; Hoehn, H. Fanconi Anemia: causes and consequences of genetic instability. In: Volff JN, editor. *Genome and Disease*. Basel: Karger Publishers; 2006; 218–242.

Kaplan, MI; Morgan, WF. The nucleus is the target for radiation-induced chromosomal instability. *Radiation Research,* 1998, 150, 382–390.

Kapultcevich, YuG; Korogodin, VI; Petin, VG. Analysis of radiobiological reactions of yeast cells. 1. Survival curves and delayed appearance of macrocolnies. *Radiobiology*, 1972, 12, 267–271.

Karpenshif, Y; Bernstein, K. From yeast to mammals cells: recent advances in genetic control of homologous recombination. *DNA repair (Amst)*, 2012, 11, 781–788.

Klein, F; Karwan, A; Wintersberger, U. Pedigree analysis of yeast cells recovering from DNA damage allow assignment of lethal events to individual post treatment generations. *Genetics*, 1990, 124, 57–65.

Korogodin, VI. *The Problems of Postirradiation Recovery*. Moscow: Atomizdat; 1966.

Korogodin, VI; Bliznik, KM; Kapultcevich, YuG; Korogodina, VL; Korolev, VG; Mezhevaya, EV. Cascade mutagenesis: regularities and mechanisms. *Proc 2nd Int NW Timoffeev-Ressovsky Conf. Dubna*, 2007, 1, 419–447.

Kumaran, R; Yang, S-Y; Leu, J-Y. Characterization of Chromosome Stability in Diploid, Polyploid and Hybrid Yeast Cells. *PLoS ONE*, 2013, 8 (7), e68094, https://doi.org/10.1371/journal.pone.0068094.

Lee, M-S; Yu, M; Kim, K-Y; Park, G-H; Kwack, K; Kim, KP. Functional Validation of Rare Human Genetic Variants Involved in Homologous Recombination Using *Saccharomyces cerevisiae*. *PLoS ONE*, 2015, 10(5), e0124152, https://doi.org/10.1371/journal.pone.0124152.

Lewis, LK; Resnick, MA. Tying up loose ends: nonhomologous end-joining in *Saccharomyces cerevisiae*. *Mutat Res*, 2000, 451, 71–89.

Little, JB. Radiation-induced genomic instability. *Int J Radiat Biol*, 1998, 74, 663–671.

Little, JB; Nagasawa, H; Pfenning, T; Vetrovs, H. Radiation-induced genomic instability: delayed mutagenic and cytogenetic effects of X rays and alpha particles. *Radiat Res*, 1997, 148, 299–307.

Luchnik, AN; Glaser, VM; Shestakov, SV. Repair of DNA double-strand breaks requires two homologous DNA duplexes. *Mol Biol Repts*, 1977, 3, 437–442.

Mao, Z; Bozzella, M; Seluanov, A; Gorbunova, V. DNA repair by nonhomologous end joining and homologous recombination during cell cycle in human cells. *Cell Cycle,* 2008, 7, 2902–2906.

McMurray, MA; Goltschling, DE. Aging and genetic instability in yeast. *Current Opinion in Microbiology,* 2004, 7, 673–679.

Melloni, E; Marchesini, R; Emanuelli, H; Fava, G; Locati, L; Pezzoni, G. Hyperthermal effects in phototherapy with hematoporphyrin derivative sensitization. *Tumori*, 1984, 70, 321–325.

Morgan, WF; Murnane, JP. A role for genomic instability in cellular radioresistance? *Cancer and Metastasis Reviews,* 1995, 14, 49–58.

Mulla, W; Zhu, J; Li, R. Yeast: A simple model system to study complex phenomena of aneuploidy *FEMS Microbiol Rev.*, 2014, 38(2), 201–212.

Petin, VG; Kabakova, NM. RBE of densely ionizing radiation for wild-type and radiosensitive mutants of yeast. *Mutat Res*, 1981, 82, 285–294.

Petin, VG; Kapultcevich, YuG. Radiation quality and the shape of dose-effect curves at low doses of ionizing radiation for eukaryotic cells. *Math Biosci*, 2014, 252, 1–6.

Petin, VG; Kim, JK. *Synergistic Interaction and Cell Responses to Environmental Factors*. New York: Nova Sciences Publisher; 2016.

Petin, VG; Kim; JK. Liquid-holding recovery kinetics in wild-type and radiosensitive mutants of yeast cells exposed to low- and high-LET radiations. *Mutat Res*, 2005, 570, 1–8.

Petin, VG; Pereklad, OV; Nili, M; Kim, JK. Yeast cells retain a memory of their original radiation-induced insult. *J Radiat Industry*, 2008, 2, 59–64.

Petin, VG; Zhurakovskaya, GP; Komarova, LN. *Radiobiological Basis of Synergistic Interaction in Biosphere*. Moscow: GEOS; 2012.

Ponnaiya, B; Jenkins-Baker, G; Bigelow, A; Marino, S; Geard, CR. Detection of chromosomal instability in α-irradiated and bystander human fibroblasts. *Mutation Research*, 2004, 568, 41–48.

Raju, MR. *Heavy Particle Radiotherapy*. New York-London: Academic Press; 1980.

Saeki, T; Machida, I; Nakai, S. Genetic control of diploid recovery after gamma-irradiation in the yeast *Saccharomyces cerevisiae. Mutation Research*, 1980, 73, 251–265.

Sheltzer, JM; Blank, HM; Pfau, SJ; Tange, Y; George, BM; Humpton, TJ; Brito, IL; Hiraoka, Y; Niwa, O; Amon, A. Aneuploidy Drives Genomic Instability in Yeast Science. *Science*, 2011, 333(6045), 1026–1030.

Shen, Z. Genomic instability and cancer: an introduction. *J Mol Cell Biol*, 2011, 3, 1-3.

Skoneczna, A; Kaniak, A; Skoneczny, M. Genetic instability in budding and fission yeast – sources and mechanisms. *FEMS Microbiol Rev*, 2015, 39, 917–967.

Stirling, PC; Bloom, MS; Solanki-Patil, T; Smith, S; Sipahimalani, P; Li, Z; Kofoed, M; Ben-Aroya, S; Myung, K; Hieter, P. The complete spectrum of yeast chromosome instability genes identifies candidate CIN cancer genes and functional roles for ASTRA complex components. *PLoS Genetics*, 2011, 7(4), e1002057, https://doi.org/10.1371/journal.pgen.1002057.

Sugasawa, K. Xeroderma pigmentosum genes: functions inside and outside DNA repair. *Carcinogenesis,* 2008, 29, 455–465.

Tolstorukov, II; Bliznik, KM; Korogodin, VI. Mitotic instability of *Pichia pinus* diploid yeast cells. Communication 1. Spontaneous segregation. *Genetika*, 1979, 15, 2140–2147.

Yuen, KWY; Warren, CD; Chen, O; Kwok, T; Hieter, P; Spencer, FA. Systematic genome instability screens in yeast and their potential relevance to cancer. *Proceeding of the National Academy of Sciences of the USA,* 2007, 104, 3925–3930.

Zhu, J; Pavelka, N; Bradford, WD; Rancati, G; Li, R. Karyotypic Determinants of Chromosome Instability in Aneuploid Budding Yeast. *PLoS Genetics*, 2012, 8(5), e1002719, https://doi.org/10.1371/journal.pgen.1002719.

Biographical Sketch

Ekaterina S. Evstratova

Affiliation: A. Tsyb Medical Radiological Research Centre – Branch of the National Medical Research Radiological Centre of the Ministry of Health of the Russian Federation

Education: Higher Professional Education "National Nuclear Research University, Moscow Engineering and Physics Institute"

Business Address: Obninsk, Russia

Research and Professional Experience:

Researcher scientist in Biophysical Laboratory of Medical Radiological Research Center in Obninsk, Russia. I am the PhD in Radiobiology in 2015. The theme of my thesis is "Synergism and cell recovery after combined action of chemical agents with ionizing radiation or hyperthermia." Research effort is devoted to quantitative estimation of recovery parameters after combined action of ionizing radiation and chemical agents. Quantitative approach describing cell recovery from potentially lethal damage as a decrease in the effective dose have been used to estimate separately the probability of recovery per time unit and the fraction of irreversible damage which cannot be repaired at all after combined action of ionizing radiation with different chemical agents. Nowadays interest is shifting to combined treatment using ionizing radiation in cancer treatment.

Honors:

Winner of the competition of the Russian President scholarships during 2015-2017 for young scientists engaged in advanced research and development in priority scientific areas "Development of innovative medical technologies for optimizing the effects of drugs that increase the radiosensitivity of tumor cells for radiation therapy." Winner of a grant of the Russian Foundation of Fundamental Research performed by the young scientist in 2016 "The synergy and optimization of drug effects: new horizons and prospective of densely ionizing radiation" (Project Manager). Taking part in the project of the Russian Humanity Scientific Foundation "Central Russia: Past, Present and Future" - Kaluga region. Scientific awards of name V. M. Klechkovsky in 2017. Winner of the Russian Federation President's grant for state support of young Russian scientists to the work "Genetic instability and aging of cells after application of various physical factors and medicinal products used in medicine."

Publications from the Last 3 Years:

1. Evstratova E. S., Petin V. G. Regularities of the effect of chemical agents on the radiosensitivity of diploid yeast cells *Saccharomyces cerevisiae*//*Modern Mycology in Russia*. 2015. V. 4. Issue. 1. P. 63-65.
2. Petin V. G., Evstratova E. S., Babina D. D. Radiobiological basis of densely ionizing radiation application in nuclear medicine/Meeting in St. Petersburg "*Modern Problems of Genetics, Radiobiology, Radioecology, and Evolution*" Dubna. 2015. P. 44.
3. Evstratova E. S., Petin V. G. Increase in radiosensitivity of cells after inhibition of their ability to repair after potentially lethal irradiation damage//*Cell and Tissue Biology*. 2015. V. 9. No. 6. P. 462–466.
4. Petin V. G., Babina D. D., Evstratova E. S. Influence of the concentration of chemical agents on the manifestation of synergistic effects//*Collection of scientific works Laureates of regional prizes and scholarships*. Issue. 11. – Kaluga. 2015. P. 131-142.
5. Evstratova E. S., Kim J. K., Kim J.-H., Petin V. G. Influence of chemical inhibitors on cell recovery after exposure to different LET radiation. In *"Proceedings of the Korean Association for Radiation Protection (KARP),"* Jeju. 2016. P. 140-141.
6. Evstratova E. S., Pereklad O. V., Petin V. G. Dependence of radiation-induced genetic instability on ploidy of yeast cells. *Radiation and Risk*. 2016. V. 25. No. 4. P. 80-89.
7. Evstratova E. S., Kim J.-H., Lim Y. K., Kim J. K., Petin V. G. Chemical inhibition of cell recovery after irradiation with sparsely and densely ionizing radiation. *Journal of Radiation Industry*. 2016. V. 10. No. 4. P. 199–204.
8. Evstratova E. S., Pereklad O. V., Khryachkova A. V. RBE a-particles to delay the formation of colonies by irradiated yeast cells. *Radiation Biology. Radioecology*. 2016. V. 56. No. 5. P. 481-486.
9. Evstratova E. S., Vorobei O. A., Filimonova A. N. Evaluation of the mechanism of radiosensitization by chemical preparations after the action of rare- and dense-ionizing radiation. *Actual problems of the humanities and natural sciences*. 2016. No. 12. Part I. P. 37-41.

10. Khryachkova A. V., Evstratova E. S., Podobed M. Yu., Petin V. G. Genetic instability of yeast cells, induced by ionizing radiation and ultraviolet light. *Contemporary Mycology in Russia*. 2017. V. 6. P. 98-100.
11. Evstratova E. S., Petin V. G., Pronkevich M. D. Respiratory mutant and liquid holding recovery inhibition in yeast cells. *International Journal of Radiation Research*. 2017. V. 15. No. 2. P. 129–139.
12. Pronkevich M. D., Evstratova E. S., Belkina S. V., Anokhin Yu. N., Petin V. G. Comparison of effects of combined effects on yeast and mammalian cells. *Medical Radiology and Radiation Safety*. 2017. T. 62. No. 6. P. 21-27.
13. Evstratova E. S., Petin V. G., Zhurakovskaya G. P. Synergistic effects and their potential significance for the influence of natural intensities of environmental factors on cell growth. *Synergy*. 2018. V. 6. P. 1–8.

In: Ionizing Radiation
Editor: Tamar Reeve
ISBN: 978-1-53613-539-8

Chapter 4

IONIZING RADIATION DAMAGE IN GA_2O_3 AND GAN WIDE BANDGAP SEMICONDUCTORS

***Stephen J. Pearton*[1,*]*, Jiangcheng Yang*[2]*, Fan Ren*[2]*, Gwangseok Yang*[3] *and Jihyun Kim*[3]**

[1]Department of Materials Science and Engineering, University of Florida, Gainesville, FL, US
[2]Department of Chemical Engineering, University of Florida, Gainesville, FL, US
[3]Department of Chemical and Biological Engineering, Korea University, Seoul, Korea

ABSTRACT

We review data on the radiation resistance of AlGaN/GaN and InAlN/GaN High Electron Mobility Transistors (HEMTs) as well as

---

* Corresponding Author Email: spear@mse.ufl.edu.

emerging Ga_2O_3 photodetectors and rectifiers to different types of ionizing radiation. Both of these wide bandgap semiconductor (nitride-based and gallium oxide-based) materials are much more radiation-hard than GaAs or Si and this is largely a result of their high bond strengths. The states in the bandgap introduced by ionizing or non-ionizing radiation and comparative figures-of-merit such as carrier removal rates in these materials are covered. The carrier removal rates in GaN, for example, are a function of initial carrier concentration and dose but not of dose rate or hydrogen concentration. Device simulators are able to reproduce the observed changes in HEMTs after radiation-damage. Non-ionizing radiation like neutrons leads to so-called Gossick disorder zones of extended damage and at high doses can pin the Fermi level. By contrast, ^{60}Co γ-rays produce less reduction in HEMT drain current than other types of radiation. Optical devices such as InGaN/GaN blue Light-Emitting Diodes (LEDs) irradiated with either protons (fluences near 10^{14} cm^{-2}) or electrons (fluences near 10^{16} cm^{-2}) exhibit degraded current-voltage and light output-current characteristics with increasing dose. A comparison of optical and electrical performance of the LEDs shows the former is more sensitive to change than the latter with both proton and electron irradiation. Ga_2O_3 shows a similar level of radiation resistance to GaN. The measured carrier removal rate for 10 MeV electrons in this material was 4.9 cm^{-1}.

Keywords: GaN, Ga_2O_3, radiation damage, carrier removal rates, point defects, disorder

1. Introduction

Gallium Nitride is attractive for use in power electronics used in air, land and satellite based communications and phased array radar, as well as in light emitters [1-9]. GaN-based blue light emitting diodes (LEDs) employ AlGaN and InGaN alloys with different compositions integrated into heterojunctions and quantum wells are used in energy-efficient white light sources and full-color displays. Gallium Oxide is attracting attention because of its high breakdown field of 8MV.cm^{-1} and it is also promising for power electronics and solar blind UV detectors, as well as extreme environment electronics (high temperature, high radiation, and high voltage (low power) switching [2, 10]. β-Ga_2O_3 is likely to be very radiation hard, based on its bond strength, but there is much basic information on the effects of radiation on charge collection, stoichiometric disturbances, defect creation, role of hydrogen, phase stability and the scaling of these effects at small volumes that is missing. There is a need to measure the effects of total dose proton, electron, gamma ray and neutron fluxes on Ga_2O_3, which has exceptionally high breakdown fields and great promise for high power, high temperature electronics [11-13].

Monoclinic β-phase Ga_2O_3 has outstanding potential for power electronics with a large direct bandgap of ~4.6 eV and the commercial availability of high quality, large diameter bulk crystals and epitaxial layers with a range of controllable n-type doping levels. Both of these materials (nitride-based and gallium oxide-based) will be subject to radiation environments if used in high temperature electronics or detectors used for military or nuclear-industry monitoring [14-27]. If used in low earth orbit satellites there will be proton irradiation of these materials, as well as neutrons or gamma rays if used in radiation-hard electronics for nuclear or military systems. Each of these forms of radiation produces different types of damage. In addition, primary defects may recombine, form complexes with each other, with dopants and with extended defects; at high energies the energy of the primary recoils becomes so high that they produce collision cascades and form heavily disordered regions with a very high defect density in the core [18-20]. In general, proton and electron irradiation in wide bandgap semiconductors produces simple point defects in a semiconductor lattice, while neutron irradiation creates extended defects called Gossick zones, which are heavily disordered core regions surrounded by a space charge region with strong band bending [14]. The response to gamma irradiation is often quite complicated. Compton electrons induced from γ-radiation create electron-hole pairs, thus changing occupancy of traps. For proton and electron damage, the device degradation scales with dose and is correlated with the nuclear or non-ionizing energy loss component of the ions traversing the active regions of the device which creates lattice displacements. The carrier removal rates for neutron damage are smaller than for proton damage [19, 20, 22]. In space-based applications, electronic systems are exposed to fluxes of protons, α-particles, and heavier ions. Additional sources of radiation include the van Allen belts, with high proton and electron fluxes and the Sun, which produces both protons and heavier ions with variable composition during solar flares. Finally, primary cosmic rays create secondary cascades of protons, neutrons, mesons and nuclear fragments whose intensity is a maximum at 18km and decreases with distance down to sea level. At normal aircraft cruising altitudes, the radiation flux encountered is several hundred times the ground level intensity [28].

In this chapter we review data for radiation damage in both Ga_2O_3 and GaN. The carrier removal rates for different types of radiation are discussed, as well as degradation of different types of devices based on these materials.

2. Radiation Damage in Ga_2O_3

The large bandgap of Ga_2O_3 allows high temperature operation and the large critical field allows high voltage operation (relative to maximum breakdown) [2-4]. Additionally, Ga_2O_3 has a high saturation electron velocity (v_{sat} = 2 x 10^7 cm/s), which is partially accountable for the high current density, I_{max} ($I_{max} \approx qn_sv_{sat}$ where q = 1.6 x 10^{-19} C, n_s = sheet charge density, v_s = electron saturation. velocity), and potentially high operating frequency as $f_t \approx v_{sat}/L_{eff}$. The Ga_2O_3 has lattice constants of a = 12.23Å, b = 3.04 Å, c = 5.80 Å and two types of two kinds of coordination for Ga^{3+} ions in the structure, namely distorted tetrahedral and octahedral, meaning the oxide ions are in a distorted cubic closest packing arrangement, and the gallium (III) ions in the two types of sites have Ga-O bond distances of 1.83 and 2.00 Å, respectively [4].

Figure 1 shows a Reflection electron energy loss spectroscopy (REELS) spectrum employed to measure the bandgap Ga_2O_3. The bandgap of the Ga_2O_3 was determined to be 4.6 ± 0.3 eV. The band alignment at SiO_2/Ga_2O_3 and $HfSiO_4/Ga_2O_3$ heterojunctions was obtained from XPS measurements and in both cases were found to be nested gap (type I) band offsets. For SiO_2 the valence band offset was 1.2eV and the conduction band offset was 2.9 eV, while for $HfSiO_4$ the valence band offset was 0.02 eV and the conduction band offset was 2.38 eV. The conduction band offsets in both cases are large and provide excellent electron confinement, but the valence band offset for $HfSiO_4$ is too small for limiting hole transport. A comparison of the literature to date on band offsets for dielectrics on Ga_2O_3 shows that SiO_2 is the best choice.

Figure 2 shows a comparison of the theoretical breakdown voltage versus bandgap for different semiconductor materials, as well as the expected on-state resistance of switching devices as a function of breakdown voltage in these systems [2, 4]. This shows the expected advantages of Ga_2O_3 over both SiC and GaN for high power switching electronics. While diamond is also very promising, the economics of large area diamond substrates are impractical. Gallium Oxide is already commercially available in large diameter wafer form, with low defect density. Table 1 shows a compilation of properties of the different semiconductor materials [1, 29]. The biggest current drawback with gallium oxide is its low thermal conductivity.

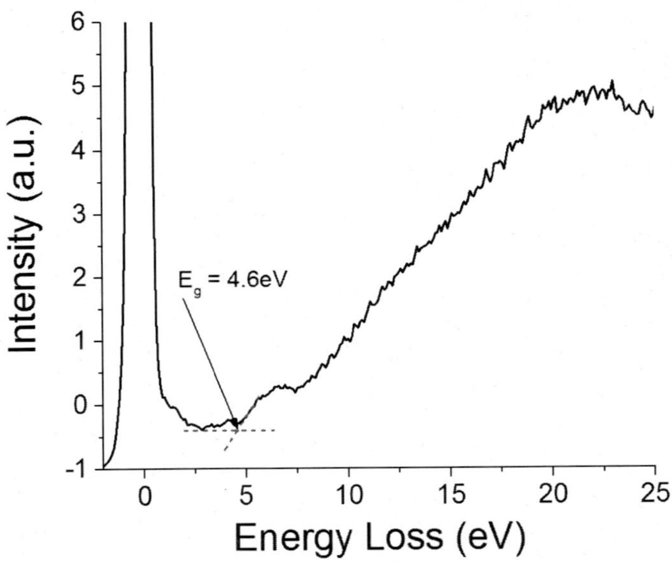

Figure 1. (top) Reflection electron energy loss spectra to determine the bandgap of bulk Ga_2O_3. Reprinted with permission from Hays et al., [44], Copyright 2017, American Institute of Physics.

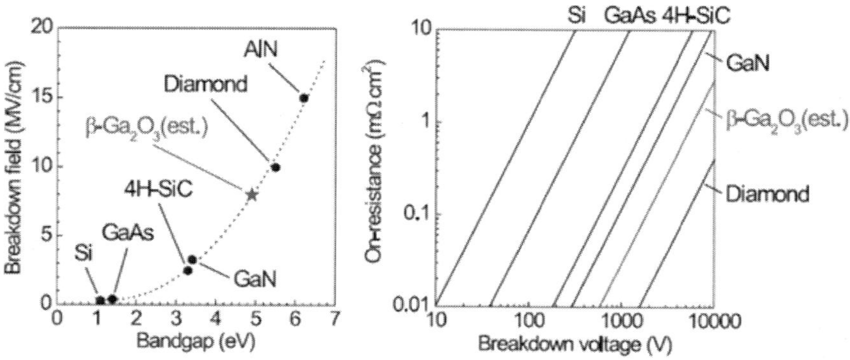

Figure 2. Breakdown voltage (left) and on-resistance versus breakdown voltage (right) for different semiconductor materials. Reprinted with permission from Mastro et al., [44], Copyright 2017, The Electrochemical Society.

Table 1. Compilation of properties of different semiconductor materials

	Si	GaAs	4H-SiC	GaN	Ga_2O_3	Notes
Bandgap E_g [eV]	1.12	1.42	3.25	3.4	4.85	
Dielectric Constant, ε	11.8	12.9	9.7	9	10	
Breakdown Field, E_c [MV/cm]	0.3	0.4	2.5	3.3	8	
Electron Mobility, μ [cm^2/V·s]	1500	8500	1000	1250	300	
Maximum Velocity, Vs [10^7 cm/s]	1	1	2	3	2	1.8 <0 0 1> and <0 1 0>, 2.0 <0 1 0>
Thermal Conductivity, λ, [W/cm·K]	1.5	0.5	4.9	2.3	0.23	0.13 W/cm·K <1 0 0>, 0.23 W/cm·K <0 1 0>
Figure of Merits / relative to Si						
Johnson = $E_c^2 \cdot V_s^2/4\pi^2$	1	1.8	277.8	1089.0	**2844.4**	power-frequency capability
Baliga = $\varepsilon \cdot \mu \cdot E_c^3$	1	14.7	317.1	846.0	**3214.1**	specific on-resistance in (vertical) drift region
Combined = $\lambda \cdot \varepsilon \cdot \mu \cdot V_s \cdot E_c^2$	1	3.7	248.6	353.8	37.0	combined power/frequency/voltage
Baliga High Frequency = $\mu \cdot E_c^2$	1	10.1	46.3	100.8	142.2	measure of switching losses
Keyes = $\lambda \cdot [(c \cdot V_s)/(4\pi \cdot \varepsilon)]^{1/2}$	1	0.3	3.6	1.8	0.2	thermal capability for power density / speed

We used vertical rectifiers fabricated on epi Ga_2O_3 on bulk β-Ga_2O_3 as test vehicle for studying the effect of electron irradiation on this material. In brief, these rectifiers were fabricated by depositing a full area back Ohmic contact of Ti/Au (20 nm/80 nm) by electron-beam evaporation. The front sides were patterned by lift-off of E-beam deposited Schottky contacts Ni/Au (20 nm/80 nm) on the epitaxial layers. The diode diameters were 100-200μm. These devices were subject to 1.5 MeV electron irradiation at fluences from 1.79×10^{15} to 1.43×10^{16} cm^{-2} at a fixed beam current of 10^{-3}A. The electron irradiation caused a reduction in carrier concentration in the epi Ga_2O_3, with a carrier removal rate of 4.9 cm^{-1}. The 2kT region of the forward current-voltage characteristics increased due to electron-induced damage, with an increase in diode ideality factor of ~8% at the highest fluence and a more than 2 order of magnitude increase in on-state resistance, as shown in Figure 3. There was a significant reduction in reverse bias current, which scaled with electron fluence. The reverse bias I-V characteristics showed the effects of carrier removal by introduction of electron traps. Figure 4 shows the I-Vs before and after the three different electron fluences. Note the strong reduction in reverse current with electron fluence when plotted on a linear scale. The on/off ratio when switching from +1 V forward bias to -10V reverse bias voltage was severely degraded by electron irradiation, decreasing from $\sim 10^7$ in the reference diodes to $\sim 2\times10^4$ for the 1.43×10^{16} cm^{-2} fluence. The reverse recovery characteristics showed little change even at the highest fluence, with values in the range 21-25 nsec for all rectifiers. These are the times needed to recover when switching from +5V to -5V.

As yet, there are no clear studies of the defect levels introduced by any form of radiation exposure in Ga_2O_3. Arehart et al., [29] found introduction of a deep state at E_c1.88 eV in neutron-irradiated samples which is thought to be related to oxygen vacancies. Yang et al., [30] reported the changes in forward and reverse I-V characteristics of rectifiers irradiated with high energy electrons.

The introduction of trapping centers is also clear from the increase in diode ideality factor and on-state resistance with increasing electron fluence, as shown in Figure 5. The ideality factor of the unirradiated control rectifiers was 1.07, consistent with thermionic emission being the dominant current transport mechanism [24-30], and an increase in this with electron irradiation is due to the introduction of generation-recombination centers. If the latter dominates the current transport, the ideality factor will be 2 [27, 28]. If the effective carrier concentration in the epitaxial layer is decreased by the introduction of electron traps, then R_{ON} will increase [29-

31]. The on-state resistance in the unirradiated diodes was 6 mΩ.cm^{-2} and this increased by more than two orders of magnitude at the highest electron fluence. The Ga_2O_3 shows comparable radiation hardness to GaN for electron irradiation, based on the similarity of carrier removal rate.

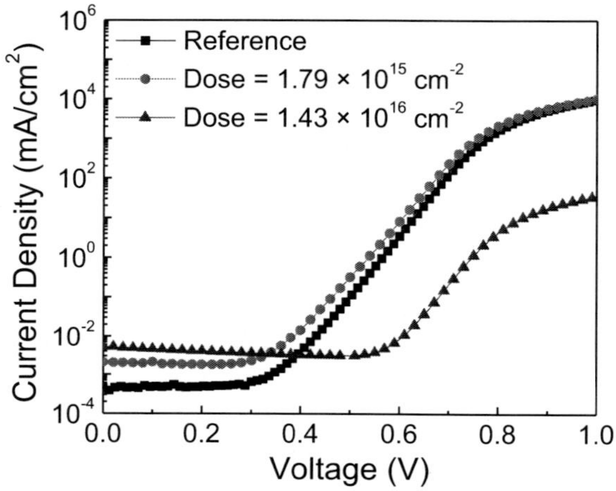

Figure 3. Forward current density-voltage characteristics before and after electron irradiation with the fluence 1.79x10^{15} or 1.43 x10^{16} cm^{-2}. Reprinted from Yang et al., [46]. Copyright 2017, American Institute of Physics.

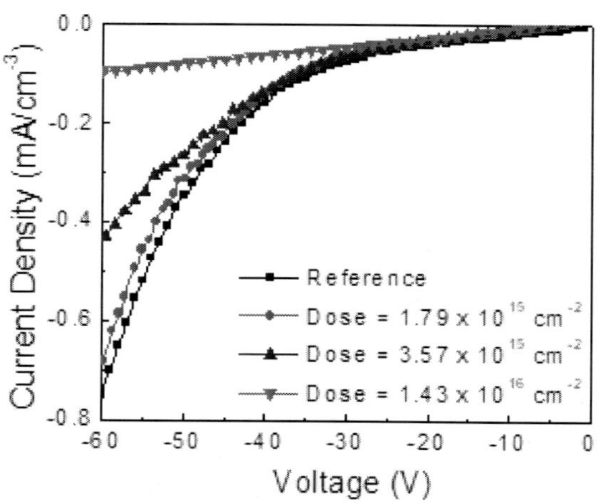

Figure 4. Reverse current density-voltage characteristics of Ga_2O_3 rectifiers before and after electron irradiation at three different fluences. Reprinted from Yang et al., [46]. Copyright 2017, American Institute of Physics.

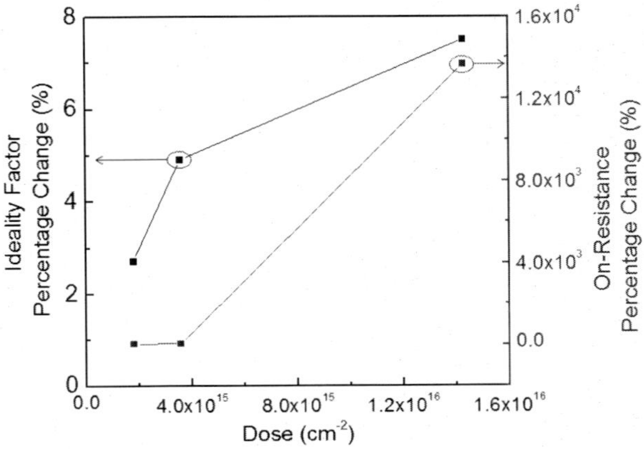

Figure 5. On-state resistance and diode ideality factor as a function of 1.5 MeV electron fluence.

Planar thin film β-Ga_2O_3 photodetectors were irradiated with 5 MeV protons at doses from 10^{13} -10^{15} cm^{-2} and the resulting effects on photocurrent, responsivity, quantum efficiency and photo-to-dark current ratio (PDCR) at 254 nm wavelength were measured at both 25 and 150 °C. These devices photoconductors were fabricated on epitaxial layers of 1 μm thick β-Ga_2O_3 on c-plane Al_2O_3 substrates. The samples were implanted with 30 keV Si ions at a dose of 10^{15} cm^{-2} and annealed 900°C for 60 s to produce a sheet resistance of 164.1 kΩ/square, much lower than that of the grown material (> 10^7 Ω/square). The contact structure was fabricated by lithographically patterning Ti/Au (20 nm/80 nm) deposited by electron-beam evaporation. The outer circle diameter of the contact was 240 μm and the inner circle diameter 220 μm while the gap between the two was 10 μm. The photocurrent increased with dose due to the introduction of damage from non-ionizing energy loss by the protons. The total calculated vacancy concentration increased from $5x10^{15}$-$5x10^{17}$ cm^{-3} over the dose range investigated [19]. The dark current increased in proportion with the implant dose, leading to a decrease in the ratio of photocurrent to dark current, as shown in Figure 6. The increase in photocurrent relative to the dark current measured under exposure to this light can be explained by the presence of defect levels within the bandgap [9, 25, 26]. We also saw an increase in the dark current of the irradiated photodetectors by illuminating with sub-bandgap (red or green laser light) for the proton irradiated samples, but not for the un-irradiated control devices. This is also consistent with our previous results (15), where we saw that the response to 254 and 365 nm photons had two components (fast and slow), originating from band-to-band transition and band-to-deep-level transitions. When the radiation energy was below band-gap energy, the current still increased due to the presence of defect levels in the band gap. In this case, the generation of photocurrent is due to a transition between the valence or conduction band and impurity or defect levels within the band gap. The photocurrent induced by 254 nm illumination increased with dose, from ~0.3 x10^{-7} A at 25 °C for a dose of 10^{13} cm^{-2} to ~10^{-6} A at a dose of 10^{15} cm^{-2} at a fixed light intensity of 760 μW/cm^2. The photo-to-dark current ratio decreased from ~60 in the control samples to ~9 after proton doses of 10^{15} cm^{-2}, with corresponding external quantum efficiencies of ~10^3 % in control samples, ~2 x10^3 % for a dose of 10^{13} cm^{-2} and 10^4 % for a dose of 10^{15} cm^{-2}. The mechanism for the increase in photocurrent is introduction of gap states since the dark current of the photodetectors was increased by illuminating with sub-bandgap (red or green laser light) for the proton irradiated samples.

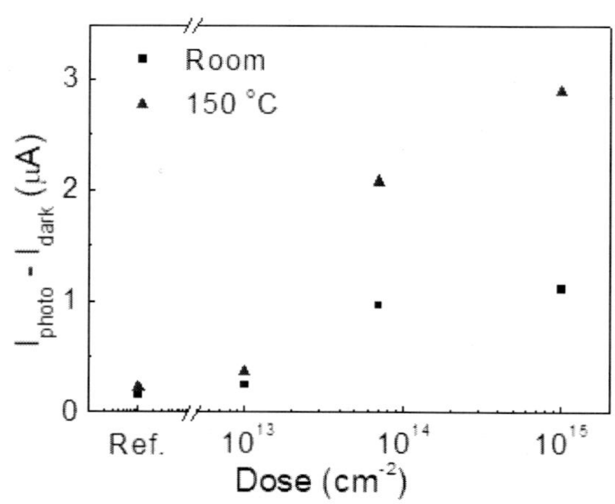

Figure 6. Difference in photo current and dark current as function of irradiation doses at room temperature and 150 °C for 254 nm. Reprinted from Ahn et al., [23]. Copyright 2017, American Institute of Physics.

In summary, Ga_2O_3 photodetectors were subject to fluences of 5MeV protons from 10^{13}-10^{15} cm^{-2}. The non-ionizing energy loss of the protons as they traverse the Ga_2O_3 layers creates states in the gap that increase the photocurrent but decrease the photo-to-dark current ratio as the proton fluence increases. The devices are still operational after the highest fluence exposure and show the promise of this material system for solar-blind photodetection in harsh temperature and radiation environments.

Proton damage in back-gated field-effect transistors (FETs) fabricated on exfoliated quasi-two-dimensional β-Ga_2O_3 nanobelts were studied by exposure to 10-MeV protons. The proton radiation damaged FETs showed a decrease of 73% in the field-effect mobility and a positive shift of threshold voltage after proton irradiation at a fluence of 2×10^{15} cm^{-2}, which corresponds to approximately 10^5 times the intensity of a solar proton event. The on/off ratio of the exfoliated β-Ga_2O_3 FETs was maintained even after proton doses of up to 2×10^{15} cm^{-2}. The radiation-induced damage in β-Ga_2O_3-based FETs was significantly recovered after rapid thermal annealing at 500 °C.

2.1. Radiation Damage in GaN

The GaInAlN materials system has a number of prominent device applications in UV and visible lighting, full color displays and high voltage electronics. Nitride-based light-emitting diodes (LEDs) and laser diodes (LDs) are the main semiconductor light sources for the UV-blue regions and are commonly employed in applications such as traffic lights, white light illumination and data storage. In addition, AlGaN/GaN photodetectors are effective in solar-blind UV detection. More recently, the use of nitride-based electronic devices has attracted interest. AlGaN/GaN and InAlN/GaN High Electron Mobility Transistors (HEMTs) and GaN or AlGaN Schottky diodes show promise for high power switching applications [1, 6]. One of the main applications for GaN electronics is radar for military and communications systems. Modern radar and satellite communications operate primarily in the microwave radio regions at frequency bands known as X (8-12 GHz), Ku (12-18 GHz), K (18-27 GHz) and Ka (27-40 GHz) bands. The ability to achieve high power efficiency and compact size are critical for electronics intended for ship, aircraft and satellite based applications. AlGaN/GaN and InAlN/GaN High Electron Mobility Transistors (HEMTs) are well suited to these military electronics needs, with device operating frequencies well within RF bands listed above. The strong bonding in binary and ternary nitrides gives them an intrinsically high radiation resistance [31-35]. The response of GaN to radiation damage is a function of radiation type, dose, dose rate, carrier density, impurity content and dislocation density. The extended defects can trap point defects created by the radiation exposure and this affects the carrier removal rate. Thus, since the dislocation density depends on growth technique and conditions, the carrier removal rate is a function of the crystal growth method. There are three common techniques used to grow GaN layers, namely Metal Organic Chemical Vapor Deposition (MOCVD), Hydride Vapor Phase Epitaxy (HVPE) and Epitaxial Lateral Overgrowth (ELOG), each with its own characteristics in terms of defect density and distribution [36-40]. One of the fundamental aspects of the response of GaN to fluxes of radiation is the mechanism for the observed changes of electrical properties of irradiated material and the position of the Fermi level after high doses of radiation. The calculation of non-ionizing energy loss (NIEL) for different types of radiation is a useful guide to how much displacement damage will be created [34]. NIEL is a calculation of the rate of energy loss due to atomic displacements as an incident particle traverses a material. There are a number of transport codes for calculating this, including the Monte Carlo code Stopping and Range of Ions in Matter (SRIM). This has traditionally been used to calculate projected ranges of ions implanted in semiconductors but also includes the stopping powers and nuclear energy loss for ions traversing a target material. The product of the NIEL and the particle fluence gives the displacement damage energy along the track.

Native defects are present in high concentrations in GaN and strongly influence its electrical and optical properties. The isolated native defects are Ga or N vacancies, interstitials, and antisites (i.e., Ga_N or N_{Ga}) but these may interact or form combinations with impurities or dopants to form more complex defects. Schematics of simple vacancies are shown in Figure 7. The Ga and N-site vacancies (V_{Ga} and V_N) exist in multiple charge states in GaN, with associated energy levels in the gap. Some studies indicate that the gallium vacancy is the most common native defect in n-type material, while the nitrogen vacancy is the most common in p -type GaN [41-43].

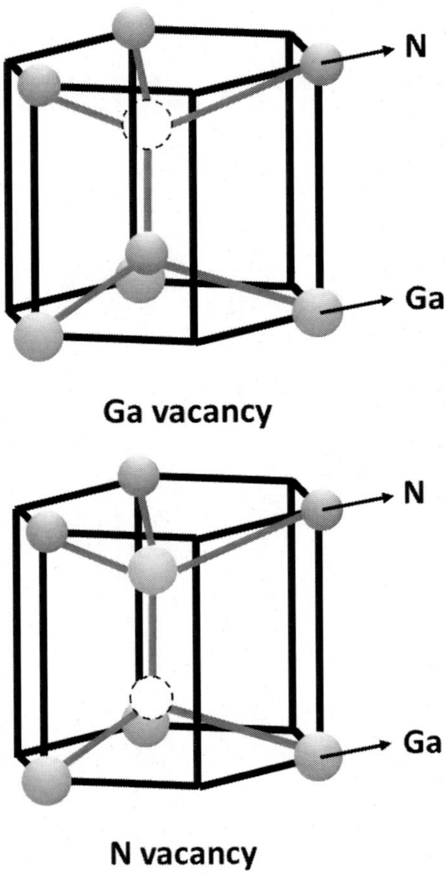

Figure 7. Schematic of a simple Ga vacancy (left) or an isolated N vacancy (right).

A simple first–order model for the defects created by ionizing radiation such as protons and electrons involves creation of four levels in the gap, namely, V_{Ga}^- with an acceptor level at E_V+1.0 eV, N_I^- with an acceptor level at E_C-1.0 eV, Ga_I^{++} with a double donor level at E_C-0.8 eV and V_N^+ with a donor level at E_C-0.06 eV [19, 20, 22]. It is important to note that literature reports often cite slightly different energies for these defects, or state them in a different convention. For example, the V_N defect may also be quoted as having an energy level of E_V + 3.28 eV in p-type GaN, which corresponds to the E_C -0.06 eV value quoted above, although others suggest E_C-0.13 eV as the energy level. Similarly, the V_{Ga} defect may be listed as having an energy level of E_C-2.5 eV, which is equivalent to the E_V+1.0 eV value discussed earlier.

Carrier removal rates by radiation exposure of GaN are found to be a strong function of initial carrier concentration, dose rate and dislocation density in the material [22]. For example, high dose rates may lead to more extensive defect complexing, or sample self-heating with consequent higher rates of self-annealing of defects. It is important to ensure that carrier removal rates are being measured in a regime in which saturation of the carrier trapping has not already occurred, leading to artificially low extracted carrier removal rates. This will obviously occur at lower doses for lightly doped samples. Figure 8 shows a compilation of removal rates in n-and p-type GaN for different types of radiation as a function of energy.

InGaN MQW LEDs with emission wavelengths from 410-525 nm irradiated with 40 MeV protons to doses of $5x10^9$-$5x10^{10}cm^{-2}$ (he highest dose is equivalent to more than 100 years in low-earth orbit) showed decreases in EL intensity by 15-25% and the reverse breakdown voltage increased by 1-2 V from their control values of ~21-29 V The results were similar for all the wavelengths investigated. The percentage change in breakdown voltage and EL intensity was independent of the initial emission wavelength over the range investigated.

For double heterostructure blue GaN/InGaN LEDs, degradation of the light output started after a dose of 10^{14} neutrons. For proton irradiated AlGaN/GaN QW LEDs the threshold dose for the onset of degradation was two orders of magnitude higher than for AlGaAs/GaAs QW LEDs (10^{12} cm^{-2} versus 10^{10} cm^{-2} for 3 MeV protons). Increasing the proton energy from 3 MeV to 5 MeV increased the dose necessary for the onset of light output degradation, due to a lower energy going into elastic collisions within the active region of devices. Higher proton doses were found necessary for changing the characteristics of proton-irradiated blue GaN/InGaN LEDs. For GaN/InGaN LEDs emitting in the green region of the spectrum, irradiation with protons at an energy of 2 MeV led to a decrease of 40% light output after a dose of 1.7×10^{12} cm^{-2} [20, 22].

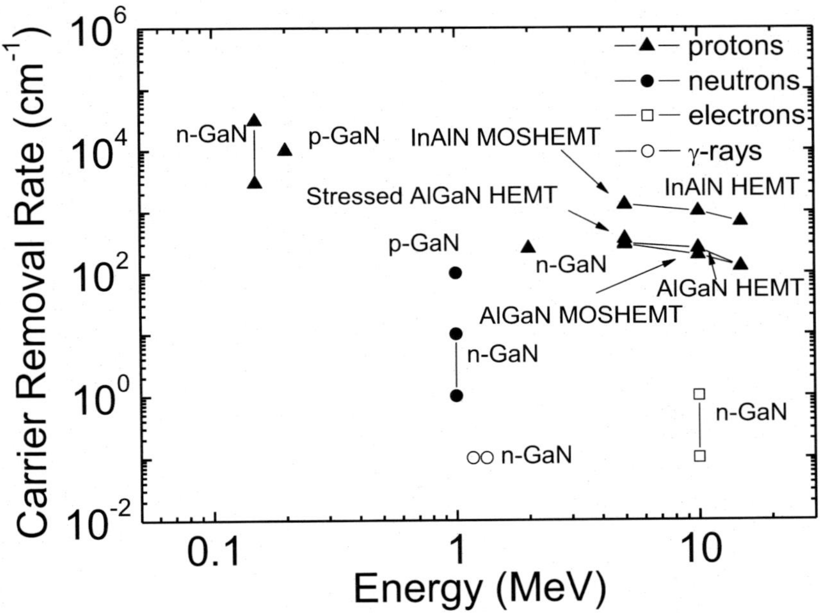

Figure 8. Carrier removal rate in single layer GaN or HEMT structures as a function of energy for different types of radiation. Reprinted with permission from Pearton et al., [22]. Copyright 2016, The Electrochemcial Society.

The degradation of properties of GaN exposed to proton or electrons is due to the introduction of point defects. For very high damage levels, there may be complexing of the radiation damage-induced defects with dopants or impurities already in the GaN. Experimental differences in the magnitude of electric fields, doping density and accuracy of the defect spectroscopy techniques all may lead to significant differences in the reported energy levels due to radiation-induced damage in [25, 26].

There have been many different energy levels introduced into both n- and p-type GaN after proton irradiation reported in the literature. These are summarized in Figure 9.

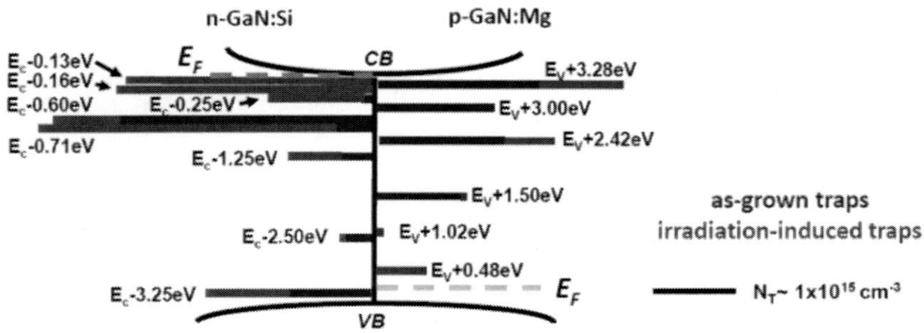

Figure 9. Schematic of position in the bandgap of defects in n- and p-GaN before and after proton irradiation (after Sasikumar et al., [25, 26]).

Radiation damage can be removed or reduced by heat treatment known as annealing. For n-GaN damaged by 2 MeV protons or 0.2-2.4 MeV electrons, measurements reveal that some of the defects begin to be removed starting at 250°C and was complete by 350 °C [20, 22]. The traps in the upper half of the bandgap associated with N interstitials annealed out at 380 °C. The Ga interstitial-related deep donors started annealing at 25 °C. By sharp contrast, the 1 eV Ga vacancy states were stable up to 500 °C. As the radiation doses are increased, the higher temperatures are needed to anneal the radiation damage. For n-GaN irradiated with very high doses of protons, the band-to-band luminescence intensity was not recovered even after annealing at 800 °C. The change of sheet resistivity of n- GaN irradiated with fast and thermal neutrons to high fluences (~10^{17} cm) recovers in a number of different stages. The resistivity initially decreased at the 150-250 °C, then went up sharply in the temperature range of 250-450°C and got smaller in the temperature range of 500-1000 °C [19]. These recovery stages correspond to reconstruction of common deep acceptors which explains the decrease in resistivity. The recovery stage at 250-450 °C is a result of complexing of interstitials to form new deep compensating centers. The final stage of recovery at 500 °C is due to V_{Ga} acceptors being removed and this leads to a recovery of the resistivity. The

Fermi level was pinned by defect states with activation energy 0.45 eV. The dominant electron traps had energy levels of 0.9 eV and 1 eV in the bandgap, which are ascribed to Gai donors and Ni acceptors, which are likely located within disordered regions. After 1000°C annealing the Fermi level was pinned near Ec-0.2 eV and the main traps were located at 0.6 eV and 0.9 eV. The density of the 0.45 eV traps pinning the Fermi level after 800 °C annealing as well as the 0.2 eV traps present for 1000°C annealing are similar and have the same concentration as the Ge donors created by neutron transmutation doping. These traps might be complexes of radiation defects with donor atoms [19, 22]. The initial conductivity was not restored even for small doses of neutrons.

The carrier removal rate in nitrides depends on radiation type, growth method and the dislocation density and distribution [20, 22]. The carrier removal rate in GaN for light particles is well accounted for by the introduction of simple defects. The carrier removal rate in nitrides depends on radiation type, growth method and the dislocation density and distribution. For fast neutrons which create large recoil cascades, carrier removal is by disordered regions in which the Fermi level in the core is pinned between the Gai donor level and the Ni acceptor level Neutron damage creates more extended damage regions. The carrier removal rate by protons, electrons and neutrons in nitride heterostructures increases in the following sequence AlN/GaN> AlGaN/GaN> InAlN/GaN. This is consistent with their average bond strengths. Proton damage has been the most widely studied and understood. Defects created by protons significantly change the breakdown voltage and off-state stressing response of both InAlN/GaN and AlGaN/GaN HEMTs [22]. The improvement was attributed to the introduction of defects. The drain-source currents of both AlGaN/GaN and InAlN/GaN HEMTs decreased 15-20% for proton irradiation with doses of 5 x10^9 cm^{-2} and 30-50% decrease at a dose of 5 x10^{10} cm^{-2}. The drain current of proton damaged HEMTs can be partially recovered (40-70%) after annealing at 300 °C.

Figure 10 shows a simple model for the main defects introduced into the bandgap of GaN during irradiation with protons or electrons, which create point defects.

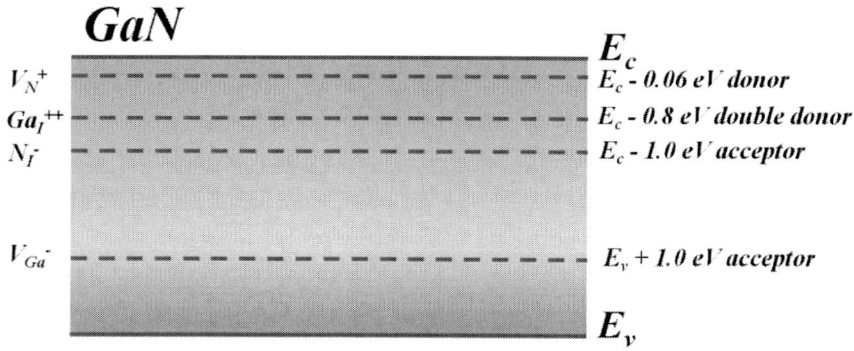

Figure 10. Main defects introduced into the bandgap of GaN by proton or electron irradiation. Reprinted with permission from Pearton et al., [20]. Copyright 2015, American Institute of Physics.

SUMMARY

The radiation hardness of GaN-based devices is about an order of magnitude higher than for their AlGaAs/GaAs counterparts, a consequence of the higher binding energy in GaN leading to a reduced introduction rate of primary radiation defects. The effect of radiation on GaN is described by a model in which the main radiation defects are due to shallow V_N and deep Ga_i donors and deep V_{Ga} and N_i acceptors. The energy levels of these are as follows: V_N donors at E_c-0.06 eV, Ga_i doubly charged donors at E_c-0.8 eV, V_{Ga} acceptors at E_v+1 eV and N_i acceptors at E_c-1 eV. In GaN, the removal of electrons and holes in the material for light particle irradiation is well-described by introduction of point defects. For energetic neutrons, there is creation of a much larger degree of lattice disorder and in that case, the removal of carriers occurs in a damaged region in which the Fermi level in the core is pinned between the Ga_i donor level and the N_i acceptor level. Radiation effects studies in Ga_2O_3 are still at a preliminary stage but this material seems to exhibit similar a similar level of radiation hardness as GaN.

ACKNOWLEDGMENTS

The project or effort depicted was sponsored by the Defense Threat Reduction Agency, HDTRA1-17-1-011, monitored by Jacob Calkins. The work at Korea University was supported by the Korea Institute of Energy Technology Evaluation and Planning (KETEP) and the Ministry of Trade, Industry & Energy (MOTIE) of Korea (No. 20172010104830) and Space Core Technology Development Program (2017M1A3A3A02015033) through the National Research Foundation of Korea funded by the Ministry of Science, ICT and Future Planning of Korea.

REFERENCES

[1] Fujita, S., Wide-bandgap semiconductor materials: For their full bloom, *Jpn. J. Appl. Phys.*, 2015, 54, 030101-1-15.http://dx.doi.org/10.7567/JJAP.54.030101.

[2] Higashiwaki, M., Sasaki, K., Murakami, H., Kumagai, Y., Koukitu, A., Kuramata, A., Masui, T. and Yamakosh, Y., Recent Progress in Ga_2O_3 Power Devices, *Semicond. Sci. Technol.*, 2016, 31, 034001-034010. https://doi.org/10.1088/0268-1242/31/3/034001.

[3] Green, A. J., Chabak, K. D., Heller, E. R., Fitch, R. C., Baldini, M., Fiedler, A., Irmscher, K., Wagner, G., Galazka, Z., Tetlak, S. E., Crespo, A., Leedy, K., and Jessen, G. H., *IEEE Electron Dev. Lett.* 2016; 37: 902-904. https://doi.org/10.1109/LED.2016.2568139.

[4] Stepanov, S. E., Nikolaev, V. I., Bougrov, V. E. and Romanov, A. E., Gallium Oxide: Properties and Applications-A Review, *Rev. Adv. Mater. Sci.,* 2016; *44:* 63-81. http://www.ipme.ru/e-journals/RAMS/no_14416/06_14416_stepanov.pdf.

[5] Millán, J., Godignon, P., Perpiñà, X., Pérez-Tomás, A. and Rebollo, J., A Survey of Wide Bandgap Power Semiconductor Devices, *IEEE Trans Power Electr.*, 2014; 29: 2155-2164. DOI: 10.1109/TPEL.2013.2268900.

[6] Bahat-Treidel, E., Hilt, O., Zhytnytska, R., Wentzel, A., Meliani, C., Würfl, J. and Tränkle, G., Fast-Switching GaN-Based Lateral Power Schottky Barrier Diodes With Low Onset Voltage and Strong Reverse Blocking, *IEEE Electr. Device Lett.* 2012; 33: 357-359. DOI: 10.1109/LED.2011.2179281.

[7] Umezawa, H., Saito, T., Tokuda, N., Ogura, M., Ri, S., Yoshikawa H and Shikata S, Leakage current analysis of diamond Schottky barrier diode, *Appl. Phys. Lett.,* 2007; 90, 073506-073508. http://doi.org/ 10.1063/1.2643374.

[8] Tadjer, M., Anderson, T. J., Feygelson, T. I., Hobart, K. D., Hite, J. K., Koehler, A. D., Wheeler, V., Pate, B. B., Eddy, C. R. and Kub, F. J., Nanocrystalline diamond capped AlGaN/GaN high electron mobility transistors via a sacrificial gate process, *Physica Status Solidi* A, 2016; 213: 893-896. DOI: 10.1002/pssa.201532570.

[9] Anderson, T. J., Hobart, K. D., Tadjer, M., Koehler, A. D., EImhoff, A., Hite, J. K., Feygelson, T. I., Pate, B. B., Eddy, C. R. and Kub, F. J., Nanocrystalline Diamond Integration with III-Nitride HEMTs, ECS *J. Solid State Sci. Technol.* 2017; 6: Q3036-3040. doi:10.1149/ 2.0071702jss.

[10] Higashiwaki, M., Sasaki, K., Kuramata, A., Masui, T. and Yamakoshi, S., Development of gallium oxide power devices, *Phys. Status Solidi* A, 2014; 211: 21-30. DOI: 10.1002/pssa.201470201.

[11] Baliga, B. J., SiC Power Devices (World Scientific, Singapore, 2005); B. J. Baliga, Power semiconductor device figure of merit for high-frequency applications, *IEEE Electron Dev. Lett.*, 1989;10, 455 -457. DOI: 10.1109/55.43098.

[12] Sze, S. M. and Ng, K. K., *Physics of Semiconductor Devices* (Wiley, Hoboken, 2007).

[13] Konishi, K., Goto, K., Murakami, H., Kumagai, Y., Kuramata, A., Yamakoshi, S., and Higashiwaki, M., 1-kV vertical Ga2O3 field-plated Schottky barrier diodes, *Appl. Phys. Lett.*2017; 110: 103506-103508. http://doi.org/10.1063/1.4977857.

[14] Polyakov, A. Y., Smirnov, N. B., Govorkov, A. V., Kolin, N. G., Merkurisov, D. I., Boiko, V. M., Korulin, A. V. and Pearton, S. J.,

Neutron transmutation doping effects in GaN, *J. Vac. Sci. Technol.* B 2010; 28, 608-612. http://doi.org/10.1116/1.3431083.

[15] White, B. D., Bataiev, M., Gross, S. H., Hu, X., Karmarkar, A., Fleetwood, D. M., Schrimpf, R. D., Schaff, W. J. and Brillson, L. J., Electrical, spectral, and chemical properties of 1.8 MeV proton irradiated AlGaN/GaN HEMT structures as a function of proton fluence, *IEEE Trans. Nucl. Sci.* 2003; 50: 1934-1940 DOI: 10.1109/TNS.2003.821827.

[16] Karmarkar, A. P., Jun, B., Fleetwood, D. M., Schrimpf, R. D., Weller, R. A., White, B. D., Brillson, L. J. and Mishra, U. K., Proton irradiation effects on GaN-based high electron-mobility transistors with Si-doped $Al_xGa_{1-x}N$ and thick GaN cap Layers, *IEEE Trans. Nucl. Sci.* 2004; 51: 3801-3805. DOI: 10.1109/TNS.2004.839199.

[17] Hu, X., Karmarkar, A., Bongim, J., Fleetwood, D. M., Schrimpf, R. D., Geil, R. D., Weller, R. A., White, R. D., Bataiev, M., Brillson, L. and Mishra, U. K., Proton-irradiation effects on AlGaN/AlN/GaN high electron mobility transistors, *IEEE Trans Nucl. Sci.,* 2003; 50: 1791-1795. DOI: 10.1109/TNS.2003.820792.

[18] Luo, B., Johnson, J. W., F. Ren, Alums, K. K., Abernathy, C. R., Pearton, S. J., Dabiran, A. M., Wowchak, A. M., Polley, C. J., Chow, P. P., Shoenfeld D and Baca A G, Influence of ^{60}Co γ-rays on dc performance of AlGaN/GaN high electron mobility transistors, *Appl. Phys. Lett.* 2002; 80: 604-606. http://doi.org/10.1063/1. 1445809.

[19] Polyakov, A. Y., Pearton, S. J., Frenzer, P., Ren, F., Liu, L. and Kim, J., Radiation effects in GaN materials and devices, *J. Mater. Chem.* C, 2013; 1: 877-885. DOI: 10.1039/C2TC00039C.

[20] Pearton, S. J., Deist, R., Ren, F., Liu, L., Polyakov, A. Y. and Kim, J., Review of radiation damage in GaN-based materials and devices, *J. Vac. Sci. Technol.* A 2013; 31: 050801-050814. http://doi.org/ 10.1116/1.4799504.

[21] Anderson, T. J., Koehler, A. D., Greenlee, J. D., Weaver, B. D., Mastro, M. A., Hite, J. K., Eddy, C. R., Kub, F. J. and Hobart, K. D., Substrate-Dependent Effects on the Response of AlGaN/GaN

HEMTs to 2-MeV Proton Irradiation, *IEEE Electron Dev. Lett.*,2014; 35, 826-828. DOI: 10.1109/LED.2014.2331001.

[22] Pearton, S. J., Ren, F., Patrick, E., Law, M. E. and Polyakov, A. Y., Review—Ionizing Radiation Damage Effects on GaN Devices, ECS *J. Solid State Science and Technol.* 2016; **5**: Q35-53. doi:10.1149 /2.0251602jss.

[23] Ahn, S., Kim, Byung-Jae, Lin, Yi-Hsuan, Ren, Fan, Pearton, S. J., Yang, G., Kim, Jihyun and Kravchenko, I. I., Effect of 5 MeV proton irradiation damage on performance of β-Ga_2O_3 photodetectors. *J. Vac. Sci. Technol.* B, 2016; 34, 051202-051206. http://doi.org/ 10.1116/1. 4950872.

[24] Kim, B.-J., Ahn, S., Ren, F., Pearton, S. J., Yang, G. and Kim, J., Effects of proton irradiation and thermal annealing on off-state step-stressed AlGaN/GaN high electron mobility transistors *J. Vac. Sci. T echnol.* B, 2016; 34, 041231-041235. http://doi.org/10.1116/1. 4959028.

[25] Sasikumar, A., Arehart, A. R., Martin-Horcajo, S., Romero, M. F., Pei, Y., Brown, D., Recht, F., di Forte-Poisson, M. A., Calle, F., Tadjer, M. J., Keller, S. P., DenBaars, S. P., Mishra, U. K., and Ringel, S. A., Direct comparison of traps in InAlN/GaN and AlGaN/GaN high electron mobility transistors using constant drain current deep level transient spectroscopy, *Appl. Phys. Lett.*,2013; 103: 033509-033511. http:// doi.org/ 10. 1063/1.4813862.

[26] Sasikumar, A., Arehart, A. R., Kaun, S. W., Chen, P., Zhang, E. X., Fleetwood, D. M., Schrimpf, R. D., Speck, J. S. and Ringel, S. A., Defects in GaN based transistors, *Proc. SPIE* 8986, Gallium Nitride Materials and Devices IX 89861C, 1 (2014).

[27] Luo, B., Kim, J., Ren, F., Gillespie, J. K., Fitch, R. C., Sewell, J., Dettmer, R., Via, G., Crespo, A., Jenkins, T., Gila, B., Onstine, A., Allums, K., Abernathy, C. R., Pearton, S. J., Dwivedi, R., Fogarty, T. and Wilkins, R., Electrical characteristics of proton-irradiated Sc_2O_3 passivated AlGaN/GaN high electron mobility transistors, *Appl. Phys. Lett.* 2003; 82: 1428-1430. http://doi.org/10.1063/1.1559631.

[28] Kang, B. S., Kim, S., Ren, F., Johnson, J. W., Therrien, R., Rajagopal, P., Roberts, J., Piner, E., Linthicum, K. J., Chu, S. N. G., Baik, K., Gila, B. P., Abernathy, C. R. and Pearton S J, Pressure-induced changes in the conductivity of AlGaN⁄GaNAlGaN⁄GaN high-electron mobility-transistor membranes, *Appl. Phys. Lett.* 2004; 85: 2962-2964 http://doi.org/10.1063/1.1800282.

[29] Arehart, A. R., Farzana, E., Blue, T. E. and Ringle S.A., presented at *2nd International Workshop on Ga_2O_3 and Related Materials*, September 2017, Parma, Italy.

[30] Yang, Ren F., Pearton, S. J., Yang, G., Kim, J. and Kuramata, A., 1.5 MeV Electron Irradiation Damage in β-Ga_2O_3 Vertical Rectifiers, *J. Vac. Sci. Technol.* B 2017; 35, 031208-1-5.

[31] Mehandru, R., Luo, B., Kim, J., Ren, F., Gila, B., Onstine, A. H., Abernathy, C. R., Pearton, S. J., Gotthold, D., Birkhahn, R., Peres, B., Fitch, R. C., Gillespie, J., Jenkins, T., Sewell, J., Via, D. and Crespo, A., AlGaN/GaN metal–oxide–semiconductor high electron mobility transistors using Sc_2O_3 as the gate oxide and surface passivation, *Appl. Phys. Lett.* 2003; 82: 2530-2532. http://doi.org/10.1063/1.1567051.

[32] Hwang, Y. S., Liu, L., Ren, F., Polyakov, A. Y., Smirnov, N. B., Govorkov, A. V., Kozhukhova, E. A., Kolin, N. G., Boiko, V. M., Vereyovkin, S. S., Ermakov, V., Lo, C. F., Laboutin, O., Cao, Y., Johnson, J. W., Kargin, N. I., Ryzhuk, R. V. and Pearton, S. J., Effect of electron irradiation on AlGaN/GaN and InAlN/GaN heterojunctions, *J.Vac. Sci. Technol.* B 2013; 31: 022206-022209 http://doi.org/10.1116/1. 4795210.

[33] Hwang, Y. H., Li, S., Hsieh, Y. L., Ren, F., Pearton, S. J., Patrick, E., Law, M. E. and Smith, D. J., Effect of proton irradiation on AlGaN/GaN high electron mobility transistor off-state drain breakdown voltage, *Appl. Phys. Lett.,* 2014; 104: 082106-082108. http://doi.org/10.1063/1.4866858.

[34] Weaver, B. D., Martin, P. A., Boos, J. B. and Cress, C. D., Displacement Damage Effects in AlGaN/GaN High Electron

Mobility Transistors. *IEEE Trans. Nucl. Sci.,* 2012; 59: 3077-3085. DOI: 10.1109/TNS.2012.2224371.

[35] Roy, T., Zhang, E. X., Puzyrev, Y. S., Fleetwood, D. M., Schrimpf, R. D., Choi, B. K., Hmelo, A. B. and Pantelides, S. T., Process Dependence of Proton-Induced Degradation in GaN HEMTs. *IEEE Trans. Nucl. Sci.,* 2010; 57: 3060-3065. DOI: 10.1109/ TNS.2010. 2073720.

[36] Selvaraj, L., Suzue, T. and Egawa, T., Breakdown Enhancement of AlGaN/GaN HEMTs on 4-in Silicon by Improving the GaN Quality on Thick Buffer Layers, *IEEE Electron*. Device Lett., 2009; 30: 587-589. DOI: 10.1109/LED.2009.2018288.

[37] Bin, L. and Palacios, T., High Breakdown (> 1500V) AlGaN/GaN HEMTs by Substrate-Transfer Technology, *IEEE Electron Device Lett*., 2010; 31, 951-953. DOI: 10.1109/LED.2010.2052587.

[38] Kanamura, M., Ohki, T., Kikkawa, T., Imanishi, K., Imada, T., Yamada, A. and Hara, N., Enhancement-Mode GaN MIS-HEMTs With n-GaN/i-AlN/n-GaN Triple Cap Layer and High- k Gate Dielectrics. *IEEE Electron Device Lett.* 2010; 31, 189-191. DOI: 10.1109/LED.2009.2039026.

[39] Liu, H.-Y., Chou, B.-Y., Hsu, W.-C., Lee, C.-S., Shue, J.-K., and Ho, C.-S., Enhanced AlGaN/GaN MOS-HEMT Performance by Using Hydrogen Peroxide Oxidation Technique, *IEEE Trans. Electron.* Dev.2013; 60 213-215. DOI: 10.1109/ TED. 2012. 2227325.

[40] Ye, P. D., Yang, B., Ng, K., Bude, J., Wilk, G., Halder, S. and Hwang, J., GaN metal-oxide-semiconductor high-electron-mobility-transistor with atomic layer deposited Al_2O_3 as gate dielectric, *Appl. Phys. Lett*. 2005; 86: 063501-063503. http:// doi.org/ 10.1063/1.186 1122.

[41] Polyakov, A. Y., Smirnov, N., Govorkov, A., Kozhukhova, E., Pearton, S. J., F. Ren, F., Liu, L., Johnson, J. W., Lim, W., Kolin, N., Veryokin, S. and Ermakov, V. S., Comparison of neutron irradiation effects in AlGaN/AlN/GaN, AlGaN/GaN, and InAlN/GaN heterojunctions. *J. Vac. Sci. Technol.* B.2012; 30: 061207-061211. http:// doi.org/ 10.1116/1.4766727.

[42] Puzyrev, Y., Mukherjee, S., Chen, J., Roy, T., Silvestri, M., Schrimpf, R. D., Fleetwood, D. M., Singh, J., Hinckley, J. M., Paccagnella A and Pantelides S T, *IEEE Trans. Nucl. Sci.* 2014, 61 1316-1320. DOI: 10.1109/TED.2014.2309278.

[43] Chen, J., Puzyrev, Y. S., Zhang, C. X., Zhang, E. X., McCurdy, M., Fleetwood, D. M., Schrimpf, R. D., Pantelides, S. T., Kaun, S. W., Kyle, E. C. and Speck, J. S., "Proton-induced dehydrogenation of defects in AlGaN/GaN HEMTs," *IEEE Trans. Nucl. Sci.*2013; 60: 4080-4086. DOI: 10.1109/TNS.2013.2278314.

[44] Hays, D. C., Gila, B., Pearton, S. J. and Ren, F., *Appl. Phys. Rev.* 2017; 4: 021301, doi: http://dx.doi.org/10.1063/1. 4980153.

[45] [45]. Mastro Michael A, Kuramata Akito, Calkins Jacob, Kim Jihyun, Ren Fan and Pearton S. J. Perspective—Opportunities and Future Directions for Ga_2O_3Electronic Materials and Processing, ECS *J. Solid State Sci. Technol.* 2017; 6(5): P356-P359; doi:10.1149/2.0031707jss.

[46] Yang, J. C., Ren, Fan, Pearton, Stephen J., Yang Gwangseok, Kim Jihyun and Kuramata Akito, 1.5 MeV electron irradiation damage in β-Ga_2O_3 vertical rectifiers, *J. Vac. Sci. Technol.* B, Nanotechnology and Microelectronics: Materials, Processing, Measurement, and Phenomena, 2017; 35, 031208; http://doi.org/10.1116/1.4983377.

In: Ionizing Radiation
Editor: Tamar Reeve
ISBN: 978-1-53613-539-8

Chapter 5

CO-MUTAGENESIS IN IRRADIATED HUMAN CELLS

E. A. Domina *

R. E. Kavetsky Institute of Experimental Pathology, Oncology and Radiobiology of National Academy of Sciences of Ukraine, Kiev, Ukraine

ABSTRACT

By using the test system of blood lymphocytes of healthy individuals, this chapter presents the co-mutagenic (potentiating) effect of the drugs – a calcium channel blocker verapamil and an antioxidant ascorbic acid-on the radiosensitivity of cells. It was found that when the lymphocytes are irradiated in a small dose (0.3 Gy), ascorbic acid (80.0 μg/blood) and verapamil (4.0 μg/ml blood) increase the frequency of chromosomal aberrations in comparison with the radiation effect by 75 and 62, 5% respectively. With an augmentation of the radiation dose, these drugs (depending on their concentration) also increase the frequency of chromosome aberrations. With a raise in radiosensitivity of

* Corresponding Author E-mail: edjomina@ukr.net.

lymphocytes (cell irradiation in the G_2-period of the cell cycle), the co-mutagenic effects of verapamil and ascorbic acid are more pronounced, as compared with the radiation in the G_0-period. This is the first time that the dependence of the formation of co-mutagenic effects in irradiated human cells was established: a dependence on the dose of ionizing radiation, the degree of radiosensitivity of cells, the concentration of drugs with co-mutagenic activity and the specificity of their action. This can be considered as an experimental justification for the correction of the designation of the drugs with identified co-mutagenic properties for the people that work in the sphere of action of ionizing radiation and the residents of territories contaminated with radionuclides due to the Chernobyl disaster.

Co-mutagenesis is a phenomenon that contributes to the burdening of the human genetic load and can be interpreted as a potentiation of mutagenic effects. The danger of co-mutagens lies in the fact that, without showing mutagenic activity, they can enhance the effects of chemical mutagens [1-4]. As early as 1962, Taylor J.H. et al. found that the inhibitor of DNA synthesis 5-fluorodeoxyuridine (FDU) increases the yield of chromosome aberrations induced by irradiation in the beans' roots [5]. In 1978, the radiobiologist and geneticist Luchnik N.V. introduced the term *pseudo-mutagen* to denote chemical agents that increase the frequency of mutations, not by inducing new chromosomal lesions, but by inhibiting the repair of hidden lesions, which in its absence would be repaired. Under these conditions, the change in the concentration of FDU did not affect the output of chromosome aberrations in human lymphocytes and indicated the pseudomutagenic properties of the drug. The scientific statement has been formulated, according to which the drugs with pseudo-mutagenic properties inhibit the repair of primary lesions in irradiated cells without causing new mutational changes.

Until now, the problem of co-mutagenesis has been scarcely studied. Research in this direction is rather fragmentary, and the results obtained are contradictory. Such a state of the problem of co-mutagenesis can be partly explained by the fact that without having its own mutagenic activity, the drugs with co-mutagenic properties "escape" (are not detected) during genotoxic screening. At the same time, they are capable, as mentioned above, of substantially potentiating the effects of well-known chemical mutagens. For example, the calcium antagonist verapamil enhances the action of bleomycin; ascorbic acid – of hydrogen hydroperoxide, bleomycin; caffeine – of methotrexate, etc. The uncontrolled presence of co-mutagens in the human habitat can increase the negative effects of the industrial, medicinal and other mutagens with which it contacts. There is no practical data on experimental work, the purpose of which would be to identify and study the features of the action of co-mutagens. In general, an information on co-mutagens was obtained as a by-product when searching for and studying the characteristics of the effect of antimutagens. Often the results attesting the co-mutagenesis are regarded as artifacts. In this regard, the development of methodological approaches to the detection of co-mutagens, including medical drugs, is of significant importance.

As it is well-known, ionizing radiation (IR) is a global environmental factor, whose effect even in small (supra-phonon) doses can cause instability in the human genome, the development of somatic pathology, reproductive problems, and increased carcinogenic risk.

Due to the steady growth of remote negative medical and biological consequences of the Chernobyl disaster, the research interest of specialists in the field of fundamental and clinical radiobiology and radiation medicine is focused on the in-depth study of mutagenic and carcinogenic effects of IR [7]. However, until now there has been no attempt to clarify how co-mutagens affect the patterns of the formation of the radiation-induced instability of the genome, especially in the region of action of the supra-phonon levels of IR. At the same time, researchers emphasize the close relationship between the radiosensitivity of human cells, tissues, organs and the appearance of malignant neoplasms [8-12].

The detection of potential co-mutagenic activity of some drugs can be carried out successfully on the basis of the test-system of human peripheral blood lymphocyte (PBL) and cytogenetic analysis. This is due to the fact that blood lymphocytes belong to the class of intermitotic cells, i.e., the most sensitive human cells to IR and other mutagens; and also taking into account the oncogenic danger of an increased level of chromosome aberrations in the cell population.

The antioxidant drugs that are most commonly used in medical practice include ascorbic acid (AA), which is characterized by an ambiguous effect on human cells [13-16]. It is established that unlike in animals, the AA in the human body is not produced, and its nutritional deficiency contributes to the development of cancer of the stomach, esophagus, mouth, cervix and other localizations [13]. Two-time Nobel laureate L. Pauling put forward the idea of prevention and treatment of neoplasia with large doses of AA (10 g per day). The encouraging results were obtained. However, under a broad clinical approbation, the effect in most cases was unreliable or even absent. The attempts by researchers to succeed due to an increase in the dose of AA to 100 g per day did neither cause an intoxication of the organism nor gave any result [16]. There is an opposite opinion, according to which it is not advisable to use the vitamins, including AA, for preventive purposes to reduce carcinogenic risk. The data on antimutagenic properties of AA is not always confirmed even in methodically closely related studies.

In the view of the abovementioned, we studied the features of the modification of radiation-induced cytogenetic effects in the culture of the donor PBL under the influence of the potential co-mutagen AA (Table 1). A cell cultivation was performed according to a standard protocol [17]. In doing so, we were guided by the Helsinki Declaration of the World Medical Association (2008), which provides for an informed consent to be obtained from the donors to participate in the study. The algorithm for the study is shown in Figure 1. X-ray irradiation of PBL cultures was carried out in the G_0- and G_2- periods of the cell cycle (at 0 and 46 hours of the cell incubation respectively). The conditions of irradiation were as follows: a dose rate - 0.89 Gy/min, a range of studied doses 0.3 - 2.0 Gy. AC was introduced into the culture of PBL immediately after irradiation in the concentration range of 20.0-40.0-80.0 μg/ml of blood, which corresponded to the therapeutic concentration, and 2 and 4 times higher than it. Metaphase analysis of cells (T-lymphocytes) was carried out in the first post-radicular mitosis. At least 300 metaphases were analyzed under each observation. The value of the mitotic index was used as an indicator of the proliferative activity of lymphocytes, for which the number of the cells in the mitosis stage was determined.

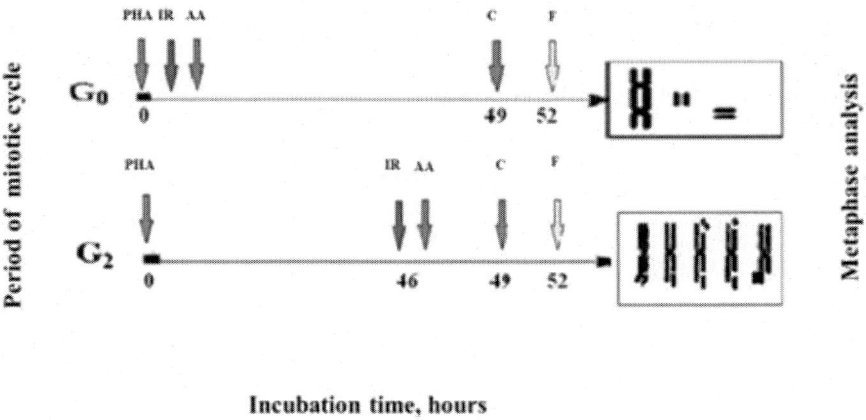

Figure 1. An algorithm for the study of the co-mutagenic properties of ascorbi acid. Designations: PHA - phytohemagglutinin (mitogen T-lymphocytes); IR - ionizing radiation, AA - ascorbic acid, C - colchicine (metaphase accumulation), F - fixation of cell culture.

This methodical approach (Figure 1) allows modeling various variants of the induced effects in combined exposure to radiation and drugs with potential co-mutagenic activity at the chromosomal level of human T-lymphocytes.

It was shown that in the range of the concentrations studied (20.0-80.0 μg/ml blood), AA does not increase the spontaneous level of chromosome aberrations in donor lymphocytes, which corresponds to the mean population values (2.0 ± 0.86 per 100 metaphases).

Analysis of the frequency of damage to chromosomes induced by irradiation of PBL in the G_0-period of the cell cycle (in the dose range of 0.3-2.0 Gy) and post-radiation exposure to AA (at a therapeutic concentration of 20.0 μg/ml of blood) has shown that the radio modifying effect of this drug is ambiguous (Figure 2.).

Under the combined effect of x-ray radiation in small doses (0.3 Gy) and AA in the therapeutic concentration (20.0 μg/ml of blood), it is observed that the overall frequency of chromosome aberrations is reduced in 1.5 times as compared to the effect of irradiation. This corresponds to the conclusions of the findings [18], according to which AA in therapeutic doses exhibits radioprotective properties by utilizing free radicals and increasing the antioxidant status of cells. When irradiating the PBL at a relatively high dose (2.0 Gy) under the influence of AA in the same therapeutic concentration, the radiation effect was potentiated, i.e., an increase in the total frequency of chromosome aberrations in 1.4 times, which indicates co-mutagenic properties of the drug (Figure 2). The observed potentiation of the radiation-induced cytogenetic effect is due to the radial markers-dicentrics (10/100 metaphases in comparison to the irradiation - 5/100 metaphase), Figure 3. Depending on the concentration of this drug (20.0, 40.0 and 80.0 μg/ml), the CMC was 35.1; 27.0 and 21.1%, respectively (Table 1).

Table 1. The frequency of radiation-induced cytogenetic effects in blood lymphocytes of healthy individuals under the influence of the combined action with AA (mean group values)

Studied groups	Frequency of aberrant cells, %	Cytogenetic indicators (for every 100 metaphases analyzed)			CMC, (%)
		Total frequency of chromosome aberrations	Frequency of chromosomal aberrations	Frequency of aberrations of chromatic type	
Control	2,2 ± 0,9	2,2 ± 0,8	0,2 ± 0,03	2,0 ± 0,6	
Control + AA (20,0 μg/ml of blood)	2,0 ± 0,6	2,0 ± 0,4	0	2,0 ± 0,9	0
Control + AA (40,0 μg/ml)	2,0 ± 0,5	2,3 ± 0,5	0,3 ± 0,1	2,0 ± 0,8	0
Control + AA (80,0 μg/ml)	2,0 ± 0,4	2,0 ± 0,4	0	2,0 ± 0,4	0
Irradiation 0,3 Gy	8,0 ± 0,7	8,0 ± 0,6	3,0 ± 0,2	5,0 ± 0,3	
0,3 Gy + AA (20,0 μg/ml)	5,0 ± 0,5	5,3 ± 0,5*	2,0 ± 0,2	3,3 ± 0,3	0
0,3 Gy + AA (40,0 μg/ml)	10,0 ± 0,3	11,2 ± 0,3	5,8 ± 0,2	6,3 ± 0,4	40,0
0,3 Gy + AA (80,0 μg/ml)	11,0 ± 0,4	13,0± 0,7*	4,0 ± 0,5	9,0 ± 0,35	62,5
Irradiation 0,5 Gy	8,3 ± 1,2	9,2 ± 0,9	5,2 ± 0,7	4,0 ± 0,1	
0,5 Gy + AA (20,0 μg/ml)	8,0 ± 1,1	8,2 ± 1,1	4,0 ± 0,8	4,2 ± 0,7	0
0,5 Gy + AA (40,0 μg/ml)	7,0 ± 0,8	7,8 ± 0,8	4,7 ± 0,7	3,1 ± 0,4	0
0,5 Gy + AA (80,0 μg/ml)	8,0 ± 0,9	8,0 ± 0,9	5,0 ± 0,4	3,0 ± 0,3	0

Studied groups	Frequency of aberrant cells, %	Cytogenetic indicators (for every 100 metaphases analyzed)			CMC, (%)
		Total frequency of chromosome aberrations	Frequency of chromosomal aberrations	Frequency of aberrations of chromatic type	
Irradiation 1,0 Gy	10,3 ± 0,8	11,3 ± 1,2	6,0 ± 0,91	5,3 ±0,8	
1,0 Gy + AA (20,0 μg/ml)	10,0 ± 0,7	11,0 ± 0,6	7,0 ± 0,6	4,0 ± 0,4	0
1,0 Gy + AA (40,0 μg/ml)	12,0 ± 0,5	12,1 ± 0,4	9,4 ± 0,5	2,7 ± 0,6	7,1
1,0 Gy + AA (80,0 μg/ml)	12,5 ± 0,4	13,3 ± 0,4	7,8 ± 0,2	5,2 ± 0,3	17,0
Irradiation 2,0 Gy	14,6 ± 1,1	17,4 ± 1,7	14,0 ± 0,8	3,4 ± 0,1	
2,0 Gy + AA (20 μg/ml)	18,0 ± 0,9	23,5± 0,9*	17,5 ± 0,6	6,0 ± 0,2	35,1
2,0 Gy + AA (40 μg/ml)	19,0 ± 1,0	22,1±0,7*	18,8 ± 0,7	3,3 ± 0,1	27,0
2,0 Гр + AA (80 μg/ml)	20,0 ± 0,8	21,4±0,9*	16,0 ± 0,5	5,4 ± 0,3	21,1

Note: * - $p < 0 < 05$ in comparison with the radiation effect.

Co-mutagenic modification coefficient (CMC) = $(M_1\text{-}M_2)/M_2 \cdot 100$, M_1 – the total frequency of radiation-induced chromosome aberrations under the influence of co-mutagens; M_2 – the total frequency of radiation-induced chromosome aberrations.

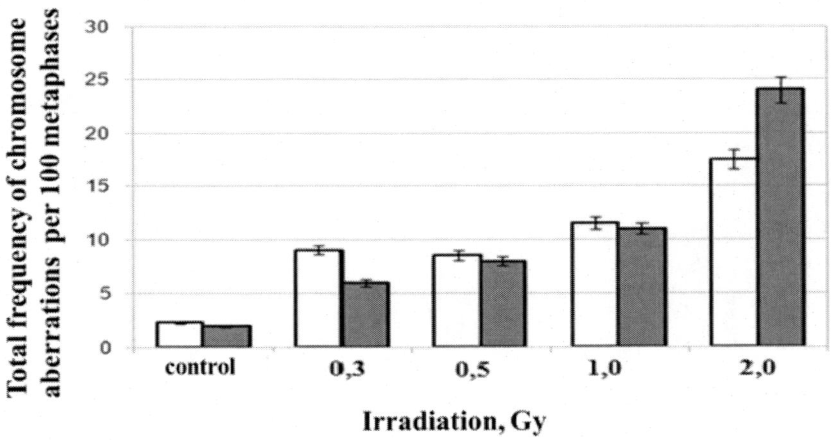

Figure 2. The total frequency of chromosome aberrations in the culture of PBL at the combined action of X-Rays and AA in a therapeutic concentration of 20.0 μg/ml blood in the G_0-period of the cell cycle. Designation: □ - irradiation of PBL in the G_0-period of the cell cycle; ■ - irradiation of PBL in the G_0-period of the cell cycle + AA 20.0 μg/ml blood.

Since the local double breaks of chromosomes are required to form exchange-type aberrations (dicentrics), the increased yield of these aberrations under the additional action of AA can be the result of primary radial damage under the influence of the drug. The effects obtained can indicate the problem of the clastogenic effect of AA on the human genome [1].

A special attention should be paid to the modification of the cytogenetic effects of small doses of ionizing radiation under the action of AA in concentrations exceeding the therapeutic value. It has been established that the additional post-radiation action of AA in concentrations of 40.0 and 80.0 μg/ml of blood increases the overall frequency of chromosome aberrations in comparison to the effect of irradiation at a dose of 0.3 Gy ~ in 1.2-1.4 times respectively (Figure 4). This also indicates the co-mutagenic activity of AA, but already in the range of concentrations exceeding in 2 and 4 times the therapeutic value.

Since the irradiation in small doses alongside with chromosome breaks can also lead to premutational potential changes in them, the additional action of co-mutagens in a range of high concentrations can promote their realization in structural rearrangements of chromosomes, including by suppressing repair enzymes [19].

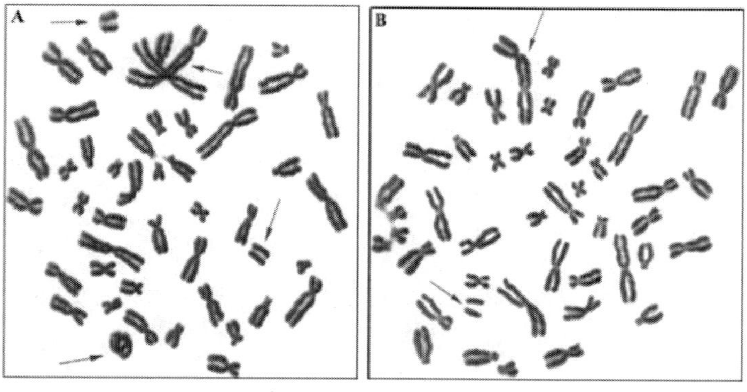

Figure 3. Aberrations of the chromosome type: A, B - exchanges and fragments.

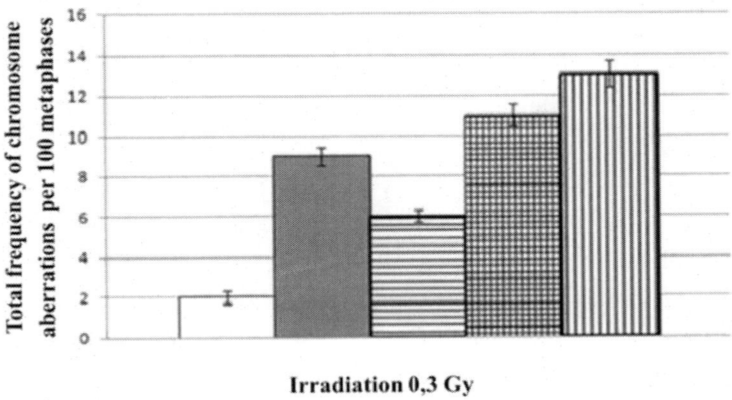

Figure 4. The total frequency of chromosome aberrations in the culture of PBL at combined action of small doses of X-Ray radiation (0.3 Gy) and AA in the concentration (20.0-40.0-80.0 μg/ml blood). Designations: □ - control, ■ - irradition, ▤ - irradiation + AA 20,0 μg/ml, ▦ - irradiation + AA 40 μg/ml, ▥ - irradiation + AA 80,0 μg/ml.

The universal indicator of the functional activity of T-lymphocytes - the ability to respond by proliferation to the action of mitogenic factors (PHA) - was evaluated. It was established that the average index of proliferative activity of stimulated blood T-lymphocytes of healthy donors was 63.7 ± 3.2‰. An independent impact of AA in the therapeutic concentration (20.0 μg/ml blood) does not change the value of the mitotic index of donor blood lymphocytes. This fact indicates that the co-mutagen AA affects neither the spontaneous level of chromosome aberrations in human immunocompetent cells nor their proliferative potential. The analysis of the proliferative activity of PBL in the conditions of the combined action of irradiation (0.3-2.0 Gy) and AA (20.0 μg/ml blood) in the G_0- period of the cell cycle has shown the following. We observe an oppression of the mitotic index of PBL under the action of small doses of IR (0.3 Gy) as compared to the control values of the indicator in about 1.5 times (40.2 ± 1.4 and 63.7 ± 3.2‰). The most pronounced suppression of the mitotic activity of cells was observed under the irradiation at a dose of 2.0 Gy, and the additional effect of AA in the therapeutic concentration, on the contrary, increased it in 2 times (exceeding even the value of intact control) (Figure 5). We believe that the observed effect is due to the "removal" of the radiation-induced block (delayed mitosis) under the influence of this drug, which reduces the time for repair of primary lesions. This is confirmed by an increase in the overall frequency of chromosome aberrations in the given experimental conditions (Figure 2).

It is a well-known fact that the when the blood lymphocytes are in the G_0-stage of the cell cycle, i.e., in the stage of rest, they are inert cells. When interacting with the mitogen, the lymphocytes are activated, which results in a number of metabolic changes that determine the response of the blast transformation. In our study, we observe the decrease in the ability of peripheral blood T-lymphocytes of healthy patients to blast transformation and, accordingly, the inhibition of their functional activity depends on the dose of radiation exposure.

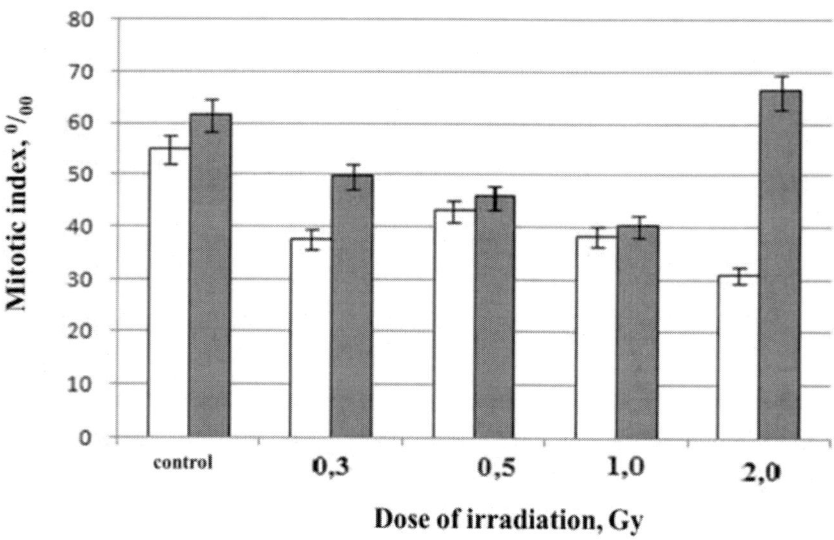

Figure 5. Combined action of irradiation (0.3-2.0 Gy) and AA (20.0 μg/ml blood) on the mitotic activity of PBL in culture. Designations: □ - irradiation of PBL in G_0-period of the cellular cycle; ■ - irradiation of PBL in the G_0-period of the cell cycle + AA 20.0 μg/ml.

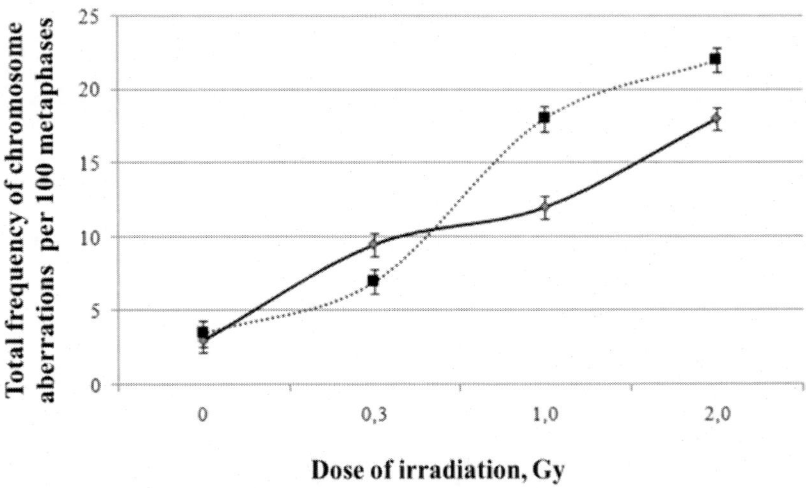

Figure 6. The total frequency of the chromosome aberrations in the culture of PBL under the combined action of X-rays and AA in the therapeutic concentration of 20.0 μg/ml of blood in the G_2-period of the cell cycle. Designations: ——◆—— - irradiation of PBL culture; ……■…… - irradiation of the culture PBL + AA 20.0 μg/ml.

The investigation of the co-mutagenic activity of this drug, depending on the radiosensitivity of the cells, represents a significant scientific interest. Consequently, we carried out a study in the most radiosensitive G_2-period of the lymphocyte cell cycle in the culture. It has been shown that during an irradiation of PBL at a dose of 0.3 Gy, the drug AA in the therapeutic concentration, as well as the experiments in the G_0- period, has a radioprotective effect (reducing the overall frequency of chromosome aberrations in 1.4 times as compared to the effect of irradiation). However, the co-mutagenic effect in G_2-period appears on the dose curve earlier in comparison with the G_0-period, starting from a dose of 1.0 Gy (12.0 ± 1.1 and 18.0 ± 1.4 respectively, that is, the potentiation of the effect in ~ 1.5 times), which is due to the high radiosensitivity of the cells in this period of the mitotic cycle (Figure 6).

Therefore, the regularities of the formation of radiation-induced structural rearrangements of chromosomes in the PBL blood of the healthy individuals under the additional modifying effect of AA have been studied. This drug, depending on the concentration when irradiating cells in small doses, can exhibit both radioprotective and co-mutagenic properties. The co-mutagenic effects of the drug are most pronounced when a somatic cell is irradiated in a relatively high dose by increasing the frequency of the radiation markers and depend on the period of the cell cycle. High concentrations of ascorbic acid potentiate the damaging effect of the small doses of IR, which is the most important practical conclusion.

An important role in the formation of the co-mutagenic effect of AA belongs to its effect on the duration of radiation-induced mitosis blocks, and thus on the proliferative potential of the cells.

The results obtained by us in the study of co-mutagenic properties of AA are generally correlated with cytogenetic data, indicating co-mutagenic properties of the drug verapamil (Vp), which is a calcium antagonist (Table 2).

It was previously established that Vp enhances the clastogenic effect of acrylamide, cyclophosphamide and dioxydine in the somatic cells of experimental animals [20, 21], increases the cytotoxic effect of some antibiotics [22], the frequency of microkernels in human blood lymphocytes when combined with erythromycin [6]. The researchers explain the mechanism of co-mutagenic action of Vp from the point of view of the "accumulation theory", according to which calcium antagonists prevent the excretion of cytotoxins from the cells, thereby increasing the potential of their mutagenic action [23. 24]. The observed increased level of chromosomal aberrations under the influence of Vp might result from a decrease in the content of intracellular calcium [25]. In accordance with the hypothesis [26], an increase in the level of intracellular calcium is associated with a DNA damage during oxidative stress. However, up to now, there was no clarification of the influence of Vp on the formation of genetic damage in irradiated human cells, particularly the immunocompetent T lymphocytes that are responsible for the antitumor protection of the body. In order to clarify this research problem, we performed a cytogenetic study on the donor BPL culture (Table 2).

The X-ray irradiation of donor blood samples (in the G_0- and G_2-period of the cell cycle) was performed at a dose power of 1.0 Gy/min, in the dose range of 0.3-2.0 Gy. A drug Vp was introduced into the culture of

PBL immediately after the irradiation in the concentration range of 1.5-2.0-4.0 μg/ml of blood, which exceeded the therapeutic value in 1.5-4 times respectively.

It was established that Vp in the range of the indicated concentrations did not influence the spontaneous level of chromosome aberrations in the lymphocytes of healthy donors, which confirms the data of the studies in which a prolonged therapy with calcium antagonists does not lead to an increase in the frequency of chromosomal changes [27, 28]. When irradiating the PBL in the dose range of 0.3-2.0 Gy and under an additional action of Vp in the investigated concentration range, the overall frequency of chromosome aberrations was increased in comparison with the radiation effect in all variants. The recorded frequency of chromosome aberrations correlated with the concentration of Vp drug (Figure 7).

Table 2. The frequency of radiation-induced cytogenetic effects in donor blood lymphocytes under the condition of combined action with verapamil (mean group values)

Studied groups	Frequency of aberrant cells, %	Cytogenetic indicators (for every 100 metaphases analyzed)			CMC (%)
		Total frequency of chromosome aberrations	Frequency of chromosomal aberrations	Frequency of chromatic aberrations	
Control	2,2 ± 0,9	2,2 ± 0,8	0,2 ± 0,1	2,0 ± 0,6	
Control + Vp (1,0 μg/ml of blood)	2,0 ± 1,3	2,0 ± 1,3	0	2,0 ± 0,7	0
Control + Vp (1,5 μg/ml)	2,3 ± 0,7	2,2 ± 0,6	0,7 ± 0,2	1,5 ± 0,3	0
Control +Vp (2,0 μg/ml)	2,0 ± 0,9	2,3 ± 0,5	0	2,3 ± 0,2	0
Control + Vp (4,0 μg/ml)	2,0 ± 0,3	2,3 ± 0,6	0	2,5 ± 0,6	0
Irradiation 0,3 Gy	8,0 ± 0,7	8,0 ± 0,6	4,0 ± 0,3	4,0 ± 0,1	
0,3 Gy + Vp (1,0 μg/ml)	8,0 ± 0,5	7,8 ± 1,0	3,3 ± 0,4	4,5 ± 0,4	0
0,3 Gy + Vp (1,5 μg/ml)	9,0 ± 0,8	10,0 ± 0,4*	6,2 ± 0,7*	3,8 ± 0,4	25
0,3 Gy + Vp (2,0 μg/ml)	8,0 ± 1,1	9,0 ± 0,8	5,6 ± 0,5*	4,4 ± 0,7	12,5
0,3 Gy + Vp (4,0 μg/ml)	10,0 ± 0,2*	14,0 ± 0,2*	9,0 ± 0,4*	5,0 ± 0,3	75
Irradiation 0,5 Gy	8,3 ± 0,4	9,2 ± 0,6	5,2 ± 0,7	4,0 ± 0,1	
0,5 Gy + Vp (1,0 μg/ml)	9,0 ± 0,3	9,0 ± 0,3	4,2 ± 0,6	4,8 ± 0,7	0
0,5 Gy + Vp (1,5 μg/ml)	8,0 ± 0,5	8,6 ± 0,9	3,3 ± 0,1*	5,3 ± 0,4*	0
0,5 Gy + Vp (2,0 μg/ml)	10,0 ± 0,4	11,2 ± 0,2*	7,2 ± 0,3*	4,0 ± 0,7	21
0,5 Gy + Vp (4,0 μg/ml)	13,9 ± 0,3*	14,5 ± 0,3*	9,3 ± 0,5*	5,2 ± 0,5*	57

Table 2. (Continued)

Studied groups	Frequency of aberrant cells, %	Cytogenetic indicators (for every 100 metaphases analyzed)			CMC (%)
		Total frequency of chromosome aberrations	Frequency of chromosomal aberrations	Frequency of chromatic aberrations	
Irradiation 1,0 Gy	11,3 ± 0,8	12,6 ± 0,2	7,3 ± 0,4	5,3 ± 0,8	
1,0 Gy + Vp (1,0 μg/ml)	11,0 ± 0,4	11,0 ± 0,4	5,9 ± 0,4	5,1 ± 0,1	0
1,0 Gy + Vp (1,5 μg/ml)	13,0 ± 0,6	13,5 ± 0,5	5,7 ± 0,3	7,8 ± 0,4*	0
1,0 Gy + Vp (2,0 μg/ml)	15,0 ± 0,8*	15,0 ± 0,8*	9,0 ± 0,6*	6,0 ± 0,2	32,7
1,0 Gy + Vp (4,0 μg/ml)	16,0 ± 0,9*	17,2 ± 1,1*	12,0 ± 0,7	5,2 ± 0,2	52,2
Irradiation 2,0 Gy	14,6 ± 0,3	17,7 ± 0,2	13,3 ± 0,5	4,4 ± 0,1	
2,0 Gy + Vp (1,0 μg/ml)	18,0 ± 0,8*	20,0 ± 0,7*	13,0 ± 0,9	7,0 ± 0,61*	14,9
2,0 Gy + Vp (1,5 μg/ml)	16,0 ± 1,1*	24,4 ± 1,2*	20,4 ± 1,1*	4,0 ± 0,8	40,2
2,0 Gy + Vp (2,0 μg/ml)	19,0 ± 0,9*	22,0 ± 1,1*	17,4 ± 0,7*	4,6 ± 0,6	26,4
2,0 Gy + Vp (4,0 μg/ml)	22,0 ± 0,6*	26,3 ± 0,7*	22,0 ± 0,5*	4,3 ± 0,2	51,1

Note: * - p < 0 < 05 in comparison with the radiation effect.

Co-mutagenic modification coefficient (CMC) = $(M_1\text{-}M_2)/M_2 \cdot 100$, M_1 – the total frequency of radiation-induced chromosome aberrations under the influence of co-mutagens; M_2 – the total frequency of radiation-induced chromosome aberrations.

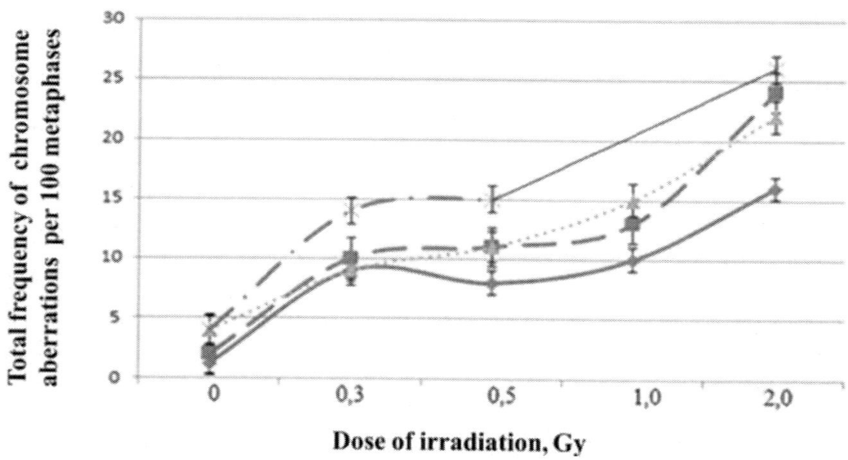

Figure 7. The total frequency of chromosome aberrations in the culture of PBL under a combined action of IR (0-2,0 Gy) and co-mutagen Vp (1.5-2.0-4.0 μg/ml blood). Designations: irradiation, irradiation + Vp (1,5 μg/ml), - irradiation + Vp (2,0 μg/ml), irradiation+ Vp (4,0 μg/ml).

The chromosome-type aberrations, whose level also depended on the concentration of co-mutagen Vp (Figure 8), were prevailing in the spectrum of lesions (Figure 3). The co-mutagenic effect of Vp was formed primarily by radiation markers- dicentric chromosomes (for example, at a dose of 0.3 Gy 7.0 dicentrics/100 metaphases, which was twice the value of this index compared to the irradiation.

The manifestation of co-mutagenic properties of Vp upon irradiation of the cells in small doses represents a particular interest (Figure 9). In the concentration range of 1.5-2.0 μg/ml blood, Vp does not significantly affect the level of radiation-induced chromosome rearrangements. However, its effect at a concentration of 4.0 μg/ml of blood (4 times the therapeutic value) potentiates the damaging effect of radiation, increasing the overall frequency of chromosome aberrations in approximately ~ 1.5 times (14.0 ± 1.12/100 metaphases). CMC = 75% (Table 2). The formation of the co-mutagenic effect of Vp also occurred, mainly, due to the chromosome-type aberrations (Figure 10).

When irradiating the PBL culture at a dose of 1.0 Gy in the most radiosensitive G_2-period of the cell cycle, the co-mutagenic effect of Vp at a concentration of 4.0 μg/ml of blood was 2.4 times higher than the effect of irradiation.

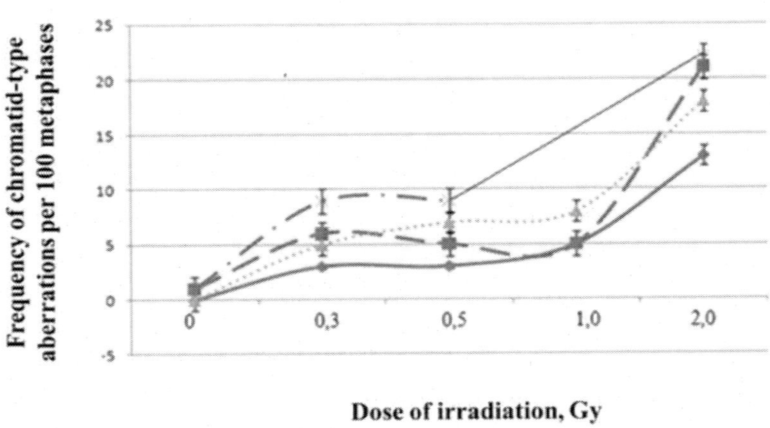

Figure 8. Frequency of chromosome-type aberrations in the donor PBL under a combined action of IR (0-2,0 Gy) and co-mutagen Vp (1.5-2.0-4.0 μg/ml blood). Designations: see Figure 7.

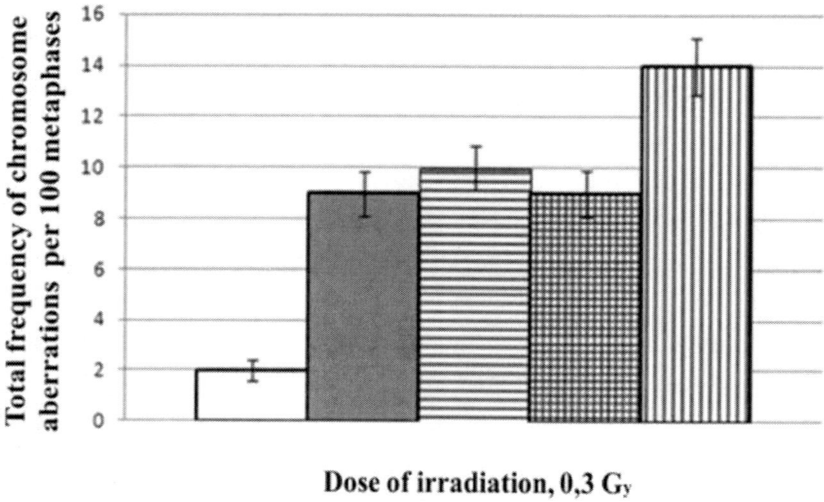

Figure 9. The total frequency of chromosome aberrations in the culture of PBL under anaction of IR in a small dose (0.3 Gy) and Vp (1.5-2.0-4.0 μg/ml blood). Designations: □ - control, ▩ - irradiation (0.3 Gy), ⊟ - Vp (1,5 μg/ml), ▦ - Vp (2,0 μg/ml), ◫ - Vp (4,0 μg/ml).

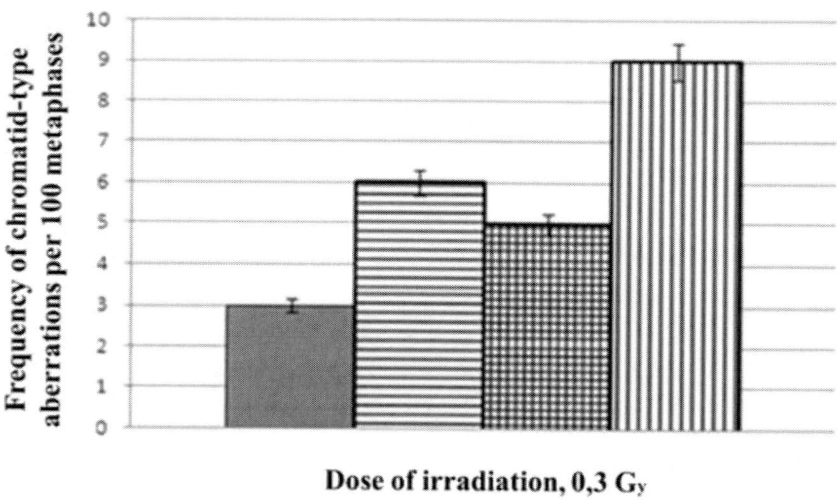

Figure 10. Frequency of chromosome-type aberrations in PBL under the combined action of IR (0.3 Gy) and Vp (1.5-2.0-4.0 μg/ml blood). Designations: see Figure 9.

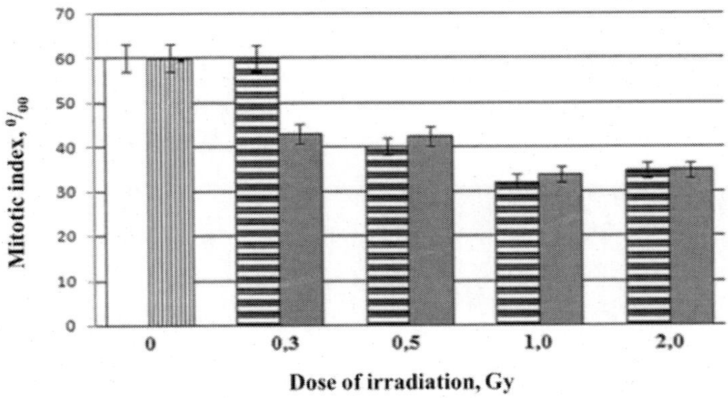

Figure 11. Mitotic activity of the donor PBL under conditions of the combined irradiation *in vitro* (0-2.0 Gy) and Vp (1.5 μg/ml blood). Designations: ☐ - control, ▥ control + Vp (1,5 μg/ml); ▤ - irradiation; ■ - irradiation + Vp (1,5 μg/ml).

It was found that Vp at a concentration of 1.5 μg/ml of blood does not affect the mitotic activity of lymphocytes (Figure 11). However, under the combined effects of IR and Vp, an inhibition of the proliferative potential of cells is observed: at a low dose (0.3 Gy) by 30% as compared to the radiation effect and intact control. Further, with an increase in the irradiation dose (including 2.0 Gy), Vp did not affect the mitotic potential of irradiated cells.

There are several points of view in radiobiology on the mechanisms involved in the formation of effects under the action of chemical agents on the irradiated cells. First, chemical agents inhibit repair processes associated with the radiation-induced DNA damage. Secondly, the degree of penetration of chemical compounds into irradiated cells may increase. Thirdly, the action of chemical agents contributes to the transformation of sublethal radiation-induced damage to potentially lethal → irreversibly lethal damage.

This indicates the absence of a universal point of view on the mechanisms of potentiation, synergism in the induction of genetic effects in conditions of the combined effects of IR and chemical agents on human cells. We do not deny the polyvariant nature of the co-mutagenesis mechanisms. However, we believe that the inhibition of the system of repair of radiation-induced damage at the genetic level of somatic cells is a dominant factor in the formation of the process. This is confirmed by the data obtained in our studies and presented in this chapter. It shows that the co-mutagenic effects in irradiated cells are formed, mainly, due to the exchanges of the chromosome type, which are the result of unrepaired critical damages - double DNA breaks.

Table 3. Radiosensitivity of the chromosomes of human blood lymphocytes under the co-mutagenic modification (Vp, AA) of radiation effects

Options studied	Regressive equation	S^2	R^2	DCF			
				0,3 Gy	0,5 Gy	1,0 Gy	2,0 Gy
IR	y = 7,4 · D + 3,5	0,23	0,92				
IR + Vp (1,0 μg/ml blood)	y = 8,1 · D + 3,7	0,35	0,95	1,03	1,02	1,15	0,89*
IR + Vp (1,5 μg/ml)	y = 10,0 · D + 3,9	0,33	0,95	0,80*	1,07	0,93*	0,73*
IR + Vp (2,0 μg/ml)	y = 9,0 · D + 4,0	0,21	0,98	0,89*	0,82*	0,84*	0,80*
IR + Vp (4,0 μg/ml)	y = 10,2 · D + 3,2	0,70	0,88	0,57*	0,63*	0,73*	0,67*
IR + AA (20,0 μg/ml)	y = 10,5 · D + 2,1	0,45	0,98	1,51	1,12	1,03	0,75*
IR + AA (40,0) μg/ml	y = 8,2 · D + 3,3	0,37	0,95	0,71*	1,10	0,93	0,79*
IR + AA (80,0 μg/ml)	y = 8,7 · D + 3,5	0,82	0,98	0,62*	1,05	0,85*	0,81*

Designation: IR – ionizing radiation; S^2 – standard error; R^2 – determination coefficient; * – co-mutagenic effect.

To analyze and generalize the results obtained, we studied the dose change factor (DCF). This is a ratio of equally effective doses in the studied and control variants, which is recommended to determine in order to evaluate the effectiveness of modifiers of radiation effects [29]. The results of the DCF calculations are presented in Table 3 and indicate the co-mutagenic nature of the effect of AA and Vp on irradiated human cells.

Based on the data in Table 3, the largest co-mutagenic effect of Vp is formed at a drug concentration of 4 μg/ml blood for the whole range of exposure doses studied, and the AA effect - at a drug concentration of 80 μg/ml blood upon irradiation in a small (0.3 Gy) dose. These effects are manifested in the potentiation of radiation effects due to the complications of chromosomal instability (increase in the number of chromosomal exchanges-beam markers). The formation of radiation-induced aberrations is related to the changes in the structure and activity of oncogenes [30]. The accumulation of chromosomal aberrations in the cell populations is a potential factor of increased carcinogenic risk [31]. Therefore, co-mutagens, which affect the radiosensitivity of human cells under irradiation, become potentially carcinogenic. The obtained data should be considered as an experimental justification for the correction of the prescription of medical drugs with identified co-mutagenic properties to the following groups of persons: (i) those working in the sphere of action of IR; and (ii) residents of the territories contaminated with radionuclides due to the Chernobyl disaster. Such measures will increase the effectiveness of primary prevention of radiogenic cancer [32].

It is of a significant interest to study the influence of medical drugs with co-mutagenic activity on the formation of genetic instability in the cells of healthy tissues of cancer patients, which fall within the zone of therapeutic irradiation. The additional effect of these drugs during the irradiation of patients can serve as a promoter of the development of secondary tumors in the long term.

CONCLUSION

It was established for the first time that the formation of co-mutagenic effects in human cells depends on the dose of radiation, the degree of radiosensitivity of the cells, the concentration of drugs with co-mutagenic activity and the specificity of their action. The most significant result of the work is the detection of the effect of co-mutagens on the radiosensitivity of blood cells of healthy individuals under the action of ionizing radiation in small doses.

REFERENCES

[1] Konopacka, M. & Rogolinski, J. (2011). Clastogenic effects in human lymphocytes exposed to low and high dose rate X-ray irradiation and vitamin C. *Nukleonika*, *56* (4), 253, 257.

[2] Domina, E. A., Pylypchuk, O. P. & Mihaylenko, V. M. (2014). Destabilization of human cell genome upon the combined effect of radiation and ascorbic acid. *Experimental Oncology*, *36* (4), 236-240.

[3] Morgan, W. F. & Sowa, M. B. (2015). Non-targeted effects induced by ionizing radiation: Mechanisms and potential impact on radiation effects. *Cancer letters*, *356* (1), 17-21.

[4] Molento, M. A., Lifschitz, J., Sallovitz., et al. (2004). Influence of verapamil on the pharmacokinetics of the antiparasitic drugs ivermectin and moxedectin in sheep. *Parasitology Research*, *92* (2), 121-127.

[5] Taylor, J. H., Haut, W. F. & Yung, J. (1962). Effects of fluorodeoxyuridine on DNA replication, chromosome breakage and reunion. *Proceeding Nat. Acad. Sci.*, *48* (2), 190-198.

[6] Grujicic, D., Milosevic-Djordjevic, O., Arsenijevic, S. & Marinkovic, D. (2008). Treatment of pregnant women with a beta mimetic and verapamil increases the micronuclei frequency in umbilical cord blood lymphocytes. *Tohoku Journal of. Experimental Medicine*, *215* (4), 363-371.

[7] *Двадцать пять лет Чернобыльской катастрофы. Безопасность будущего*. (2011). Национальный доклад Украины. Киев, Украина: KIM, 368 с. [*Twenty five years of Chernobyl disaster. The security of the future. - The National report of Ukraine* (2011). Published by KIM, Kiev, 368 p. *In Russian*].

[8] Foray, N., Colin, C. & Bourguignon, M. (2012). 100 years of individual radiosensitivity: How we have forgotten the evidence. *Radiology*, *264* (3), 627-631.

[9] Auer, S., Keller, U., Schmidt, M., et al. (2014). Individual radiosensitivity in a breast cancer collective is altered depending on a patients age. *Radiology and Oncology*, *48* (1), 80-86.

[10] Shah, D. J., Sachs, R. K. & Wilson, D. J. (2012). Radiation-induced cancer: a modern view. *British Journal of Radiology*, *85* (1020), 1166-1173.

[11] Vral, A., Willems P., Claes, K., et al. (2011). Combined effect of polymorphisms in Rad51 and Xrcc3 on breast cancer risk and chromosomal radiosensitivity. *Molecular Medicine Reports*, *4* (5), 901-912.

[12] Yang, T. C., Georgy, K. A., Mei, M., et al. (1995). Radiogenic cell transformation and carcinogenesis. *Gravitatonal and Space Research*, *8* (2), 106-112.

[13] Mirvish, S. S. (1986). Effects of vitamin C and E on N-nitro compound formation, carcinogenesis and cancer. *Cancer*, *58* (suppl.), 1842-1858.

[14] Shamberger, R. J. (1984). Genetic toxicology of ascorbic acid. *Mutation Research*, *133* (2), 135-159.

[15] Anderson, D., Basaran, V., Blowers, S. D. & Edwards, A. J. (1995). The effect on antioxidants on bleomycin treatment *in vitro* and *in vivo* genotoxicity assays. *Mutation Research*, *329* (1), 37-47.

[16] Park, C. H., Kim, W. S., Park, C., et al. (1999). Clinical disease suppression and reduction in acute myeloid leukemia and solod tumors by very high dose of α-ascorbic acid: a new concept and in scarch of molecular targets. *Clinical Cancer Research*, *5* (suppl.), 3784.

[17] *Cytogenetic Dosimetry: Applications in Preparedness for and Response to Radiation Emergencies*. (2011). Published by IAEA, Vienna, 232 p.

[18] Jagetia, G. C. (2007). Radioprotective potential of plants and herbs against the effects of ionizing radiation. *Journal of Clinical Biochemistry and Nutrition*, *40* (2), 74-81.

[19] McNeill, D. R., Heng-Kuan, Wong., Avinash, Narayana. & Wilson, D. M. (2007). Lead promotes abasic site accumulation and co-mutagenesis in mammalian cells by inhibiting the major abasic endonuclease Ape1. *Molecular Carcinogenesis*, *46* (2), 91-99.

[20] Ferguson, L. R. & Baguley, B. C. (1988). Verapamil as a co-mutagen in the Salmonella/mammalian microsome mutagenicity test. *Mutation Research Letters*, *209* (1), 57-62.

[21] Nesterova, E. V., Durnev, A. D. & Seredenin, S. B. (1999). Verapamil contributes to the clastogenic effects of acrylamide, cyclophosphamide and dioxidine of somatic cells of BALB/C and C57BL mice. *Mutation Research*, *440* (2), 171-179.

[22] Toffoli, G., Simone, F., Corona, G., et al. (1995). Structure-activity relationship of verapamil analogs and reversal of multidrug resistance. *Biochemical Pharmacology*, *50* (8), 1245-1255.

[23] Scheid, W., Weber, J., Rottgers, U. & Traut, H. (1991). Enhancement of the mutagenicity of anticancer drugs by the calcium antagonists and fenidine. *Arzneimittel-Forschung Drug Research Journal*, *51* (9), 901-904.

[24] Liu, Y. & Huang, H. (1997). Involment of calcium-dependent protein kinase C in arsenite-induced genotoxicity in Chinese hamster ovary cells. *Journal of Cellular Biochemistry*, *64* (3), 423-433.

[25] Friedman, J., Shabtai, F., Sandowski, U., et al. (1990). Ca antagonist verapamil and tumor promoter 12-O-tetrade canoylphorbol 13-acetate (TPA) induce chromosomal aberrations in human lymphocytes. *Mutation Research Letters*, *244* (2), 135-139.

[26] Halliwel, B. & Aruoma, O. I. (1991). DNA damage by oxygen-derived species. Its mechanism and measurement in mammalian system. *FEBS Letters*, *281* (1-2), 9-19.

[27] Andreassi, M. G., Picano, E. & Ry, S. D. (1999). Effects of chronic long-term therapy with calcium antagonists on cytogenetic damage in humans. *Journal of Hypertension*, *17* (6), 843-846.

[28] Milosević-Djordjević, O., Grujicić, D., Joksić, G. & Marinković, D. (2011). *In vitro* evaluation of the genotoxicity of ritodrine and verapamil in human lymphocytes. *Human and Experimental Toxicology*, *30* (5), 398-405.

[29] Гродзинский, Д. М. (2000). *Радиобиология*. Киев: Лыбидь, 448 с. [Grodzynsky, D. M. (2000). *Radiobiology*. Published by Lybid, Kiev, 448 p. *In Russian*].

[30] McConkey, E. H. (2004). *How Human Genome Works*. Published by Jones and Bartlett Publishers, Inc., Boston-Toronto-London-Singapore, 288 p.

[31] Hagmar, L., Stromberg, U., Bonassi, S., et al. (2004). Impact of types of lymphocyte chromosomal aberrations on human cancer risk: results from Nordic and Italian cohorts. *Cancer Research*, *64* (6), 2258-2263.

[32] Дёмина, Э. А. (2016). *Радиогенный рак: эпидемиология и первичная профилактика.* Киев: Наукова Думка, 196 с. [Domina, E. A. (2016). *Radiogenic cancer: epidemiology and primary prevention.* Published by Naukova Dumka, Kiev, 196 p. *In Russian*].

In: Ionizing Radiation
Editor: Tamar Reeve
ISBN: 978-1-53613-539-8

Chapter 6

WATER MOLECULES AS QUANTUM MECHANICAL SENSORS AND PRIMARY MESSENGERS FOR NON-IONIZING RADIATION EFFECTS ON CELLS

Sinerik Ayrapetyan, PhD*
UNESCO Chair in Life Sciences, Life Sciences International Postgraduate Educational Center, Yerevan, Armenia

ABSTRACT

The Membrane Theory, which is based on the principles of classical thermodynamics, explains the biological effects of environmental factors on living cells by the changes of membrane electrical conductivity. Thus, the experimental data on biological effects of extremely weak physical signals (with the intensity even less than thermal threshold), which are unable to activate the ionic channels in membrane and have non-linear dose-dependent effect on cell, cannot be explained from the point of this

* Corresponding Author Email: info@biophys.am.

theory. Therefore, the messenger(s), through which the biological effect of such weak signals on cells and organisms is realized, should have quantum-mechanical character.

It is known that the valence angle of a water molecule, determining its physicochemical properties, is extremely sensitive to weak chemical and physical factors. At present, it is also well established that cell hydration is a fundamental parameter determining cell functional activity, which is realized by regulating the activity of intracellular macromolecules by their folding-unfolding mechanism as well as by the membrane surface-dependent changes of the number of functionally active protein molecules in membrane, having enzymes, receptors and ionic channels-forming properties. Considering the fact that the cell membrane permeability for water is more than 10 times higher than for inorganic ions, it is hypothesized that, upon the effect of chemical and physical factors, the changes of net water fluxes through the membrane precede the activation of ionic fluxes in the membrane.

For supporting this hypothesis, in this chapter the experimental data of the effects of temperature, light intensity, background ionizing and non-ionizing radiation on physicochemical properties of water, water solution and cell function are presented.

Keywords: water structure, radiation, membrane, ionic channels, water fluxes, cell hydration, Na^+/K^+ pump, intracellular messengers, Na^+/Ca^{2+} exchange

INTRODUCTION

There is an asymmetric distribution of substances (including inorganic ions) between intracellular and extracellular media in living cells. It is traditionally accepted that such an asymmetric distribution of inorganic ions between both sides of cell membrane is responsible for generation of the so-called membrane potential (MP).

There are two controversial theories that explain such an asymmetric distribution of inorganic ions between extracellular and intracellular media. One of these is the Membrane Theory (MT), which considers asymmetric distribution as a result of semipermeable properties of cell surface membrane for ions [1], while the second is the Sorption Theory (ST), according to which such a distribution of ions is due to different sorption properties of intracellular structure for ions [2, 3, 4].

Although Sorption and Membrane Theories have diverse approaches on the mechanism of asymmetric distribution of inorganic ions between intracellular and extracellular media, both theories are based on classic thermodynamic laws and they consider the changes in distribution of inorganic ions between extracellular and intracellular media as messengers for signal transduction in cell. However, from this perspective it is impossible to explain the abundant experimental data on the biological effects of extremely weak signals (extremely low concentrations of chemical substances and non-ionizing radiation (NIR)), which are unable to significantly change the intracellular ionic homoestasis but can modulate cell functional activity in non-linear dose-dependent manner [5, 6, 7, 8]. Thus, it is assumed that both ST and MT, which are based on the knowledge of classic thermodynamics cannot explain the biological effect of extremely weak signals, having intensity less than the thermal threshold. Therefore, cell should have an extremely sensitive and a universal sensor with quantum-mechanical nature, through which the biological effects of weak chemical and physical signals on cells and organisms are realized. On the basis of literature and our own works of the last decade, we have developed a hypothesis, according to which water molecules (not inorganic ions) with quantum-mechanical sensitivity serve as primary messengers through which the weak chemical and physical signals are able to modulate cell metabolic activity [9, 10, 11]. The following well-documented experimental facts serve as bases for this hypothesis: a) water structure is extremely sensitive to weak physical and chemical factors [12], b) membrane permeability for water is much higher than for inorganic ions [13, 14], c) osmotic water uptake by cell is balanced by metabolic water efflux from the cells [15, 16], d) water molecules are released in cytoplasm during intracellular oxidative phosphorylation process [17], e) auto-regulation of Na^+/K^+ pump activity is induced by cell hydration [18] f) water efflux from the cells inactivates ionic inward currents through the membrane and regulates a number of functionally active ionic channels in membrane [19], g) intracellular water regulates the activity of macromolecules by folding-unfolding mechanims [20], h) intracellular signaling system, which is sensitive to cell hydration, contols $[Ca^{2+}]_i$ [21, 22] and i) the number of functionally active membrane proteins, having enzymes [18], receptors [23] and ionic channels forming properties depends on the cell surface [19]. The present chapter shortly presents the experimental data on quantum-mechanical sensitivity of water structure, which serves as a primary messenger through which the biological effect of extremely low frequencies of EMF (ELF EMF) on cells is realized.

Quantum-Mechanical Sensitivity of Water Physicochemical Properties

It is known that according to quantum mechanical calculations the valence angle in water molecules between O-H bounds should be 90°. However, in fact this valence angle is near 105° because in water the minimal repulsion of the positively charged hydrogen atoms increases the angle due to the strong polarity of the H-O bounds [24]. Due to a long hydrogen bound (0.28) in water having an electrostatic nature and a comparatively weak energy (14.2 – 20.9 kJ) the "polimer" water structure is very labile and sensitive to different environmental factors. The structure of liquid water is being continuously changed from the moment of its formation. The character of such changes depends on physical and chemical characteristics of the surrounding medium [12]. Since temperature, light intensity and background radiation (BGR) are common physical components of environmental medium, their impacts on water and the properties of physiological solution (PS) were subject of our previous study [25].

The water has polar liquid properties which slightly dissociates disproportionately into the hydroxonium ion – $H_3O^+_{(aq)}$ and associates hydroxide ions – $OH^-_{(aq)}$ according to the following equation:

$$2H_2O \longleftrightarrow H_3O^+_{(aq)} + OH^-_{(aq)}$$

Therefore, the specific electrical conductivity (SEC) of distillated water (DW) is mostly determined by these ions. Although the temperature elevation leads to the decrease of aqua medium viscosity, which brings to the increase of the water SEC, the decrease of water dielectric permeability brings to the decrease of SEC and in sum of these processes the temperature elevation leads to the increase of SEC of electrolytes [26].

As it can be seen on Figure 1, the temperature-dependency curve of DW SEC is not monotonic: in range of 0-10°C the SEC is less temperature-sensitive, while in rang of 10-40°C (the temperature-dependent part of curve) it has strong linearly rising character. In case of PS[1] the range of temperature, which is less sensitive, prolongs up to 15°C and the slope of the curve is less cranked than in case of DW. The non-linear temperature sensitivity of DW and PS SEC and diverse characters between temperature-sensitive curves of DW and PS can probably be explained by the number of temperature-dependent structural anomalies [27].

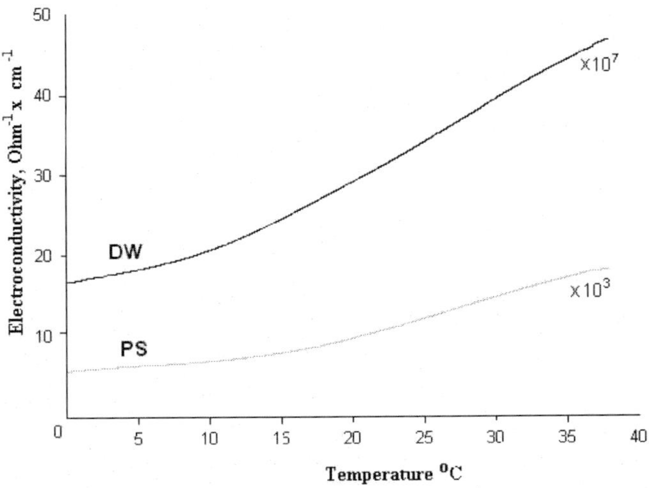

Figure 1. The temperature-dependent changes of SEC of DW (black curve) and PS (gray curve). The time duration of heating from 0 to 40°C was 30 minutes.

[1] PS contained (in mM) 137 NaCl, 5.4 KCl, 1.8 CaCl2, 1.05 MgCl2, 5 C6H12O6, 11.9 NaHCO3, and 0.42 NaH2PO4 and was adjusted to pH 7.4 with NaOH.

As the water density near 4°C is the highest and the clustered structure is less variable, it was suggested that the comparative study of water properties at 18°C and 4°C temperature will give a possibility to estimate the role of water density in determination of its sensitivity to illumination and BGR.

The data on illumination-and BGR-induced effects on PS samples, which are presented on Figure 2, show that the PS SEC at 18°C was increased only in low BGR (LBGR) (<1 μRoentgen/hour) *in Dark* (0.405%, p <0.05, n=10), while at 4°C condition this effect disappeared and slightly decreasing effect (by 0.97%, p <0.05, n=10) of normal BGR (NBGR) (17-18μRoentgen/hour) *in Dark* appeared. From these data it can be concluded that dark- and LBGR-induced effects on the PS SEC depend on water density.

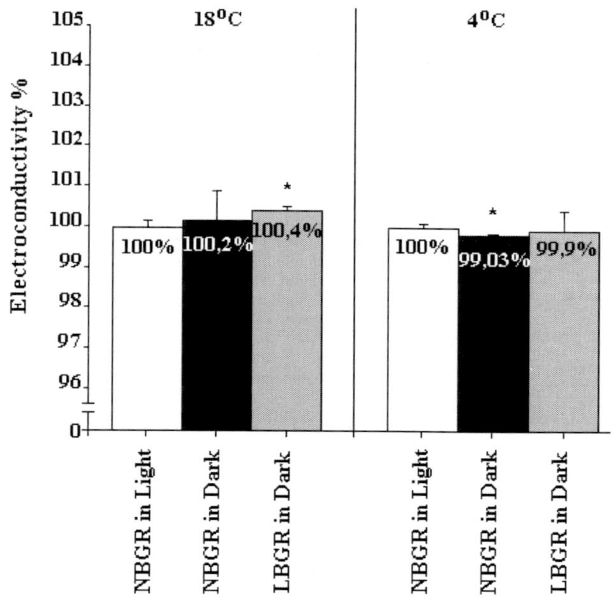

Figure 2. The effect of 10 min incubation on the SEC of PS in *NBGR in Light* (white), *NBGR in Dark* (black) and *LBGR in Dark* (gray) at 18°C and 4°C conditions. Bars indicate the mean of 10 independent measurements, vertical bars represent the standard error of the mean (SEM), $^{*}p < 0.05$ as compared with control *(NBGR in Light)*.

It is known that the heat fusion period (HFP) of aqua solution depends on its polarity. The study of HFP of frozen PS samples (10 minutes pre-incubated) in *NBGR in Light*, *NBGR in Dark* and *LBGR in Dark* at 18°C and 4°C temperature) at room temperature indicates that HFP fusion kinetics of frozen PS crystals strongly depends on temperature of incubation before freezing, BGR and illumination of pre-incubation in environmental media (Figure 3).

As there is a close negative correlation between the water heat fusion and its molecule dissociation, it allows to suggest a hypothesis that water properties (heat fusion and molecule dissociation) are also sensitive to illumination and BGR.

As it is demonstrated above, there is a strong dependency of physicochemical properties of water on environmental media, such as illumination, BGR and temperature, which can be explained by quantum-mechanical sensitivity of water structure. Thus, in the next series of experiments the dependency of the effect of ELF EMF on physicochemical properties of PS at different environmental media was studied.

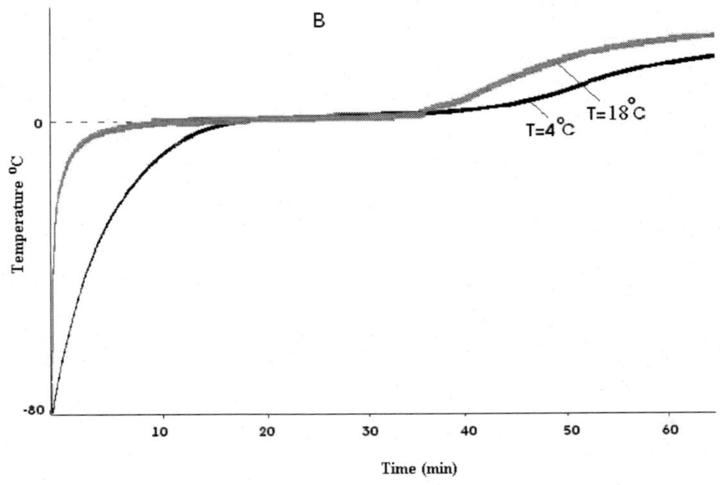

Figure 3. The effect of 10 min incubation on HFP (at room temperature) of frozen DW at 18°C and 4°C in *NBGR in Light* (white), *NBGR in Dark* (black) and *LBGR in Dark* (gray) conditions. Bars indicate the mean of 10 independent measurements, vertical bars represent the SEM, $*p < 0.05$, $**p < 0.01$ as compared with control *(NBGR in Light)*.

The Effect of ELF EMF on Physicochemical Properties of Water Depending on the Environmental Medium

At present the biological effect of NIR, such as ELF EMF can be considered as a proven fact [28]. However, the detailed mechanisms of its biological effects on cells and organisms are not clear yet. The main barrier in the study of these problems is the difficult reproducibility of the experimental data obtained in different laboratories.

From the point of the present knowledge on water structure, ELF EMF could modify the water structure by two pathways: a) by changing the valence angle in water molecules and b) by mechanical vibration (MV) of dipole molecules of water.

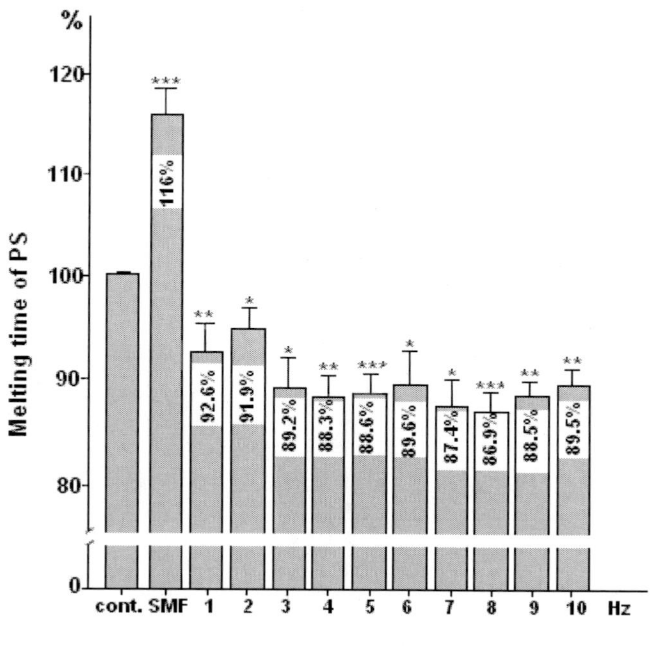

Figure 4. The frequency-dependent (0-10Hz) effect of ELF EMF on HFP of PS at NBGR in Light condition (during 10 min) at 18°C. Columns indicate the mean of 10 independent measurements, vertical bars represent the SEM, *p < 0.05, **p < 0.01, ***p <0.005 as compared to control.

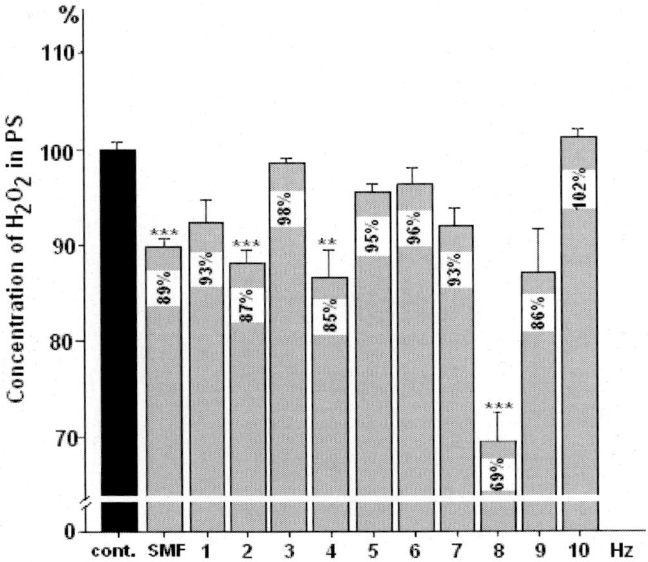

Figure 5. The frequency-dependent (0-10Hz) effect of ELF EMF on H_2O_2 content in PS at NBGR in Light condition (during 10 min) at 18ºC. SMF was chosen as a source of 0 Hz. Columns indicate the mean of 10 independent measurements, vertical bars represent the SEM, $^*p < 0.05$, $^{**}p < 0.01$ $^{***}p < 0.005$ as compared to control.

As it can be seen on Figure 4, 10-minute exposition of PS samples in static magnetic field (SMF) (0Hz) prolonged the HFP by 16%, as compared to control (non-treated sample), while 1-10Hz ELF EMF treatment strongly shortened the HFP compared to the SMF-treated one, as well as to their control samples. More pronounced effects were observed at 4Hz, 7Hz and 8Hz (by 11.7%, 12.6% and 13.1%, accordingly).

As it was noted above, the HFP serves as a marker for solution of polarization and determination of water molecule ionization, which, as a result of collision with O_2, generates the reactive oxygen species (ROS) [26] serving as strong modulators for cell metabolic activity. The results of the comparative study of the effects of 0-10 Hz ELF EMF on the H_2O_2 content of PS are shown on Figure 5.

As it can be seen on Figure 5, the treatment by SMF caused a decrease of H_2O_2 content by 11% as compared to control sample, at the same time, the implication of 1-9Hz ELF EMF also decreased the peroxide content (the most pronounced effects were recorded at 4Hz (by 15%) and 8Hz

(31%)) and only in case of 10Hz statistically non-significant elevation (by 2%) was observed.

These data clearly indicate that there are pronounced frequency "windows" of the effectsof ELF EMF on H_2O_2 content and HFP of PS, which are observed at 4Hz and 8Hz.

Taking into consideration the fact that the illumination, temperature and BGR have modulation effects on the H_2O_2 formation and the HFP of PS, it is predicted that the observed EMF effects would depend on the initial level of these factors in experimental medium.

Figure 6 and Figure 7 present the data on modulating effects of illumination, BGR on 4Hz and 8Hz ELF EMF (windows at which its effect is more pronounced) on H_2O_2 formation in PS at 18°C and 4°C temperature.

The experiments performed on the 4Hz and 8Hz ELF EMF–induced effects on H_2O_2 contents in PS treated at 4°C and 18°C for 10-minute in *NBGR in Light* (R = 17μRoentgen/hour, Ev = 500-550 lux), *NBGR in Dark* (R = 17μRoentgen/hour, Ev = 0 lux) and *LBGR in Dark* (R = <1μRoentgen/hour, Ev = 0 lux) conditions had shown that the 4Hz ELF EMF had depressing effect (by 6%) on H_2O_2 in *NBGR in Light* medium at 18°C. This effect was reversed in *NBGR in Dark* (elevation was observed by 5.5%) and disappeared in *LBGR in Dark* (Figure 6a). In case of 8Hz treatment the depression was observed in *NBGR in Light* and *LBGR in Dark* (15.5% and 13.3%, accordingly), while in *NBGR in Dark* it had no effect (Figure 6b).

The experiments executed at 4°C condition, where aqua medium had higher density, showed that the 4Hz ELF EMF treatment slightly changed the H_2O_2 content in PS, and only in *NBGR in Light* condition, the statistical verification of data was observed (by 3.8%). The 8Hz ELF EMF treatment of samples in cold medium caused a depression of peroxide content in all experimental media, but it was more expressed in *NBGR in Light* and in *NBGR in Dark* conditions by 16.2% and 13%, accordingly (Figure 7a and 7b).

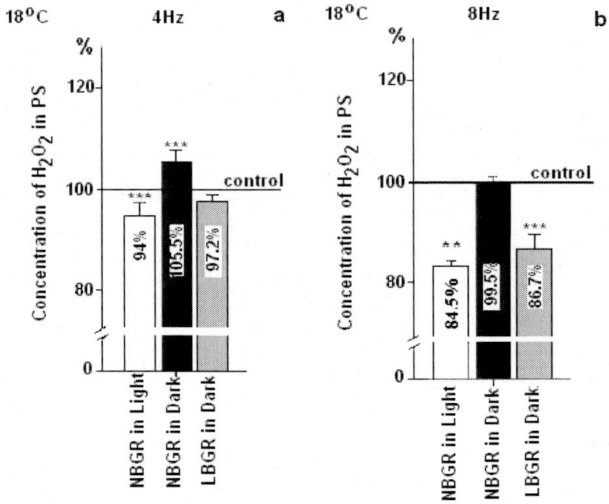

Figure 6. The 4Hz (6a) and 8Hz (6b) ELF EMF- induced effects on H_2O_2 content in PS incubated in NBGR in Light (white), NBGR in Dark (black) and LBGR in Dark (gray) conditions at 18°C. Columns indicate the mean of 10 independent measurements, vertical bars represent the SEM,*$p < 0.05$, ***$p < 0.005$ as compared to control (NBGR in Light).

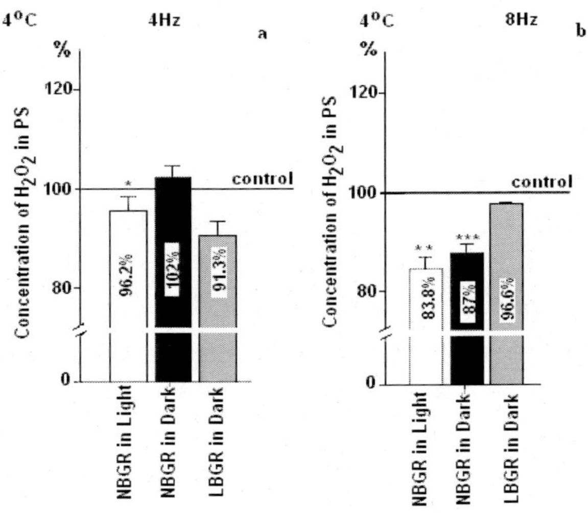

Figure 7. The 4Hz (7a) and 8Hz (7b) ELF EMF-induced effects on H_2O_2 content in PS incubated in NBGR in Light (white), NBGR in Dark (black) and LBGR in Dark (gray) conditions at 4°C. Columns indicate the mean of 10 independent measurements, vertical bars represent theSEM,*$p < 0.05$, ***$p < 0.005$ as compared to control (NBGR in Light).

WATER UPTAKE BY CELL AND ITS SENSITIVITY TO EMF

The fact that the water fluxes through the ionic channels in the membrane precede ionic currents has been elucidated by our previous study of water fluxes effect on ionic currents through the neuronal membrane. It has been shown that water fluxes through the membrane open (activate) the ionic channels in case when water fluxes has the same direction as the electrochemical gradients of ions and close (inactivate) the ionic channels when they have opposite directions [19, 29, 30].

Figure 8 shows that the transient recovery of the action potentials in 10^{-8}M tetrodotoxin (TTX)-containing solution (TTX blocks Na^+ channels in membrane) when water flow through the membrane was applied. The solutions were applied in the following order. First the axon was perfused by hypertonic external solution and an isotonic internal solution (hyper/iso) and outward water flow through the membrane was present. The addition of 10^{-8}M TTX led to complete suppression of the action potentials for 1.5-2 min. After replacement of the external hypertonic solution (at the first arrow) by isotonic one (iso/iso), outward water flow through the membrane disappeared and the transient recovery of the action potentials took place, despite the presence of TTX in the solution. Then the action potentials were blocked by TTX again. The same phenomenon took place in case of replacement of the external isotonic solution (at the second arrow) by a hypotonic solution (hypo/iso) when water flow through the membrane was in the inward direction. In all cases the axon was internally perfused by an isotonic solution and the ionic gradients were the same. i.e., the conditions differed only in the direction of the water flow through the membrane. When the above mentioned solutions were substituted in the reverse order no recovery of the action potential was observed. In the first stage when the axon was perfused by hypertonic external and isotonic internal solutions (hyper/iso) and an outward flow through the membrane was present some part of sodium ionic channels was inactivated by an outward water flow. After replacement of the external hypertonic solution by an isotonic one (at the first arrow) when outward water flow through the membrane disappeared, additional water sodium ionic channels were activated which were subsequently blocked by TTX. Then upon the replacement of the external isotonic solution by a hypotonic one at the second arrow (hypo/iso) when the water flow through the membrane became inward again, new additional ionic channels were activated which were again blocked by TTX.

These data clearly indicate that sodium channels in the neuronal membrane have heterogenic character with different thresholds of activation. Thus, it can be assumed that water influx could not only change cell hydration but also activate inward going currents.

Our previous study has shown that the SMF and 4Hz pulsing magnetic field (PMF) have depression effects on thermodynamic properties of water and water solutions because they have depression effect on the dissociation of water molecules [9]. As it can be seen on Figure 9, the 4Hz-magnetized PS has depression effect on inward Na^+ current (I_{Na}) in internally perfused squid axon bathing in 50% (250 mM) Na^+-containing sea water. It is worth to note that in normal Na^+ (500mM) containing sea water (SW) both hypertonic and magnetized solutions have no significant effects on the amplitudes of I_{Na}.

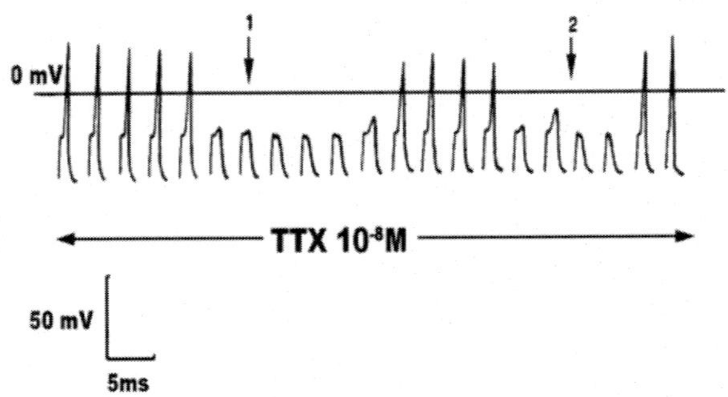

Figure 8. The transient recovery of the action potentials in a squid giant axon in 10^{-8}M TTX-containing solution at changes of the tonicity of the external solution. Originally the axon was perfused by hypertonic external and internal isotonic solutions (hyper/iso) and outward water flow through the membrane was present. At the first arrow the external hypertonic solution was replaced by an isotonic one (iso/iso) and water flow through the membrane was stopped. Then at the second arrow the external isotonic solution was replaced by a hypotonic one (hypo/iso) and inward water flow through the membrane was present. Outward and inward water flows were produced by adding or removing 500mM glucose from the external isotonic solution [19].

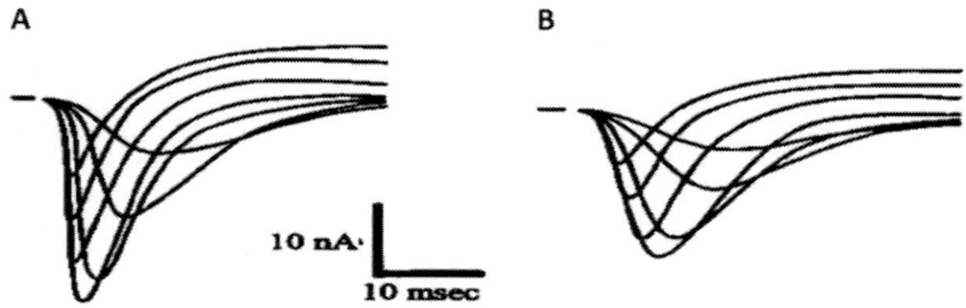

Figure 9. The effect of magnetized PS on intracellular perfused squid axon bathing in SW containing 250 mM Na^+ A) In normal SW, B) In SW preliminary exposed with 0.2mT for 30 minutes.

Our previous study on the EMF sensitivity of water uptake by plant seed in dormant and germination states has shown that simple osmotically driven water uptake by seeds in dormant state is EMF insensitive, whereas in germination period the water uptake by seeds becomes EMF sensitive: the water uptake by seeds pretreated with EMF is much higher than in case of sham-treated water [31].

Metabolic Generation of Water Efflux from the Cells and Its EMF Sensitivity

Taking into consideration that the osmotic gradient on membrane is directed from outside to inside, it is supposed that the osmotic gradient should generate water influx into the cell, which is balanced by the metabolic water efflux from the cell. The metabolic water efflux from the cell takes place due to both ion transport mechanism pushing osmotically active particles from the cells and the release of endogenous water molecules in cytoplasm during intracellular oxidative phosphorylation processes. According to Lehninger during these processes in cell 42 water molecules are released per one molecule glucose oxidation [17].

Considering the fact that the membrane permeability for water is much higher than for inorganic ions [13], it is suggested that the metabolic control of balancing the water fluxes through the membrane could be more flexible than ionic gradients on membrane, i.e., the net osmotically driven water influx precedes electrochemical ionic gradients driving ionic influxes (Na^+ and Ca^{2+}), while the net water efflux through the membrane has inactivation effect on these inward going currents.

It is known that the electrogenic Na^+/K^+ pump working in stoichiometry of $3Na^+:2K^+$ and being highly ATP utilizing machine in cells has a crucial role in generation of metabolic water efflux from the cell: from one side it pushes out more osmotically active particles (Na^+) from the cell than taking from outside the cell (K^+), from the other side it increases the release of endogenous water molecules in cytoplasm by stimulation of oxidative phosphorylation processes. Thus, the Na^+/K^+ pump not only compensates the changes of intracellular ionic composition, as it is suggested by MT, but also inhibits the membrane permeability for inward ionic currents, such as I_{Na} and Ca^{2+} currents (I_{Ca}), by generation of metabolic water efflux from the cell [19, 32]. The fact that the Na^+/K^+ pump-induced water efflux from the cells could have a double effect on ionic currents through the membrane can be easily demonstrated by the study of the effect of osmotic shock on outward potassium membrane current in intact neurons (Figure10).

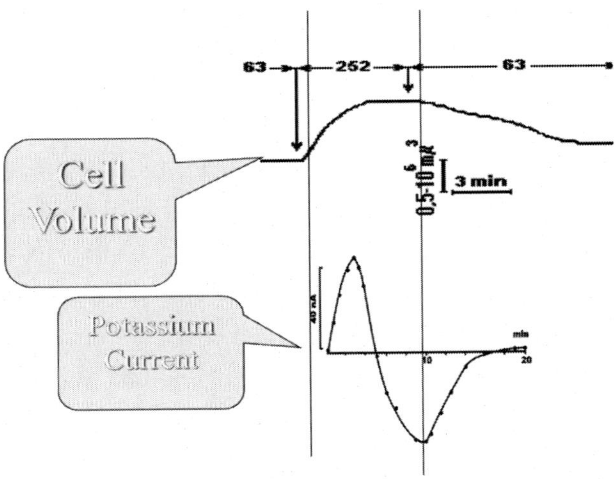

Figure 10. The time course of both cell volume (up curve) and K outward current (down curve) changes after replacing the normal isotonic solution by a hypertonic one, and the recovery of the K current after returning to the isotonic solution. The hypertonic solution was replaced by the isotonic solution after the 10th minute [19].

Figure 10 shows that the transfer of isolated snail neuron in hypertonic solution has biphasic effect on potassium currents (I_k) through the membrane: before stabilizing the cell volume in hypertonic solution the water efflux has activation effect on I_k, while after stabilization a decrease of the amplitude of this current is observed. More detailed investigation of the osmotic shock effect on cell volume has shown that water fluxes precede ionic currents in the membrane, while the cell surface shrinkage in hypertonic solution decreases the number of active ionic channels in the membrane [30].

THE ROLE OF CYCLIC NUCLEOTIDES-DEPENDENT Na^+/Ca^{2+} EXCHANGE AND ITS EMF SENSITIVITY

As it has been mentioned above, the Na^+/K^+ pump has a crucial role in generation of water efflux from the cells and its activation leads to cell shrinkage [15, 33]. Our earlier study has shown that there are three families of ouabain (specific inhibitor for Na^+/K^+-ATPase) receptors in neuronal membrane having low ($>10^{-7}$M), middle (10^{-7} to 10^{-9}M) and high ($<10^{-9}$M) affinities. Among these receptors $>10^{-7}$M ouabain has an inhibitory, while $<10^{-7}$M has activation effect on the $^{22}Na^+$ efflux from the cells [18]. Later it has been shown that low concentration ouabain-induced activation of $^{22}Na^+$ efflux from the cells is due to the cAMP-dependent activation of Na^+/Ca^{2+} exchange in reverse mode (R) [34]. At present, it is well established that the above mentioned three families of ouabain receptors correspond to three catalytic isoforms (α_1, α_2 and α_3, respectively) of Na^+/K^+-ATPase having different functional activities: α_1 and α_2 isoforms have Na^+/K^+ pump function, while α_3 isoform has only intracellular signaling function through which the $[Ca^{2+}]_i$ is regulated without changing intracellular Na^+ [35, 36, 37].

As Ca^{2+} has a key role in realization of the biological effects of EMF [38, 6] and Na^+/Ca^{2+} exchange has a central role in regulation of $[Ca^{2+}]_i$ [36], it was suggested that the cAMP/cGMP-dependence of Na^+/Ca^{2+}

exchange could serve as one of the metabolic gates through which the biological effects of extremely low concentrations of ouabain and EMF are realized. By our previous and literature data it has been shown the cAMP has activation effect on R Na^+/Ca^{2+} exchange [34], while cGMP elevates the release of Ca^{2+} from the cells by Ca^{2+}-pump and Na^+/Ca^{2+} exchange in forward mode (F) [39, 40].

As intracellular signaling systems are sensitive to various factors, it is predicted that the cAMP/cGMP-dependent Na^+/Ca^{2+} exchange could serve as a common metabolic gate through which the biological effect of low concentrations of biologically active substances and weak physical signals is realized. This suggestion was confirmed by the study of extremely low concentrations of synaptic transmitters and hormones which are unable to activate ionic channels in membrane but strongly modulate the functional properties of cell membrane through changing the activity of intracellular cyclic nucleotides-dependent Na^+/Ca^{2+} exchange [7, 39, 41].

In order to find out which of the aforementioned intracellular messenger systems (cAMP or cGMP) is sensitive to cell hydration, the effects of 2-decenoic acid (DA)-induced cell hydration on intracellular contents of cAMP and cGMP have been studied (Figure 11).

The DA, which is not metabolized by the cells, strongly increases (about 40%) membrane fluidity (lateral motion of lipids) [43]. As it can be seen on Figure 11, the DA-induced cell hydration increases cGMP content up to 124%, while cAMP level is not changed. From these data it can be concluded that cell hydration-induced activation of soluble guanylyl cyclase (sGC) serves as a hydration-sensitive metabolic gate through which cell hydration modulates intracellular metabolic activity.

Experimental conditions	cAMP	cGMP
Control	100±5	100±5
DA	100±9	124±8

Figure 11. The DA-induced cell swelling in K^+-free solution and intracellular cAMP and cGMP contents [42].

By measuring the dose-dependent effect of extremely low concentrations of synaptic transmitters (Acetylcholine (ACh) and GABA) on intracellular cyclic nucleotides it has been shown that the threshold for activation of cGMP system is more sensitive than of cAMP system [44].

The above presented data indicate that the cyclic nucleotide-dependent Na^+/Ca^{2+} exchange is extremely sensitive not only to low concentrations of ouabain but also to biologically active substances, such as ACh and GABA [7, 8, 44] and a number of synthesized ACh derivatives [45]. Taking into consideration this, it can be assumed that cyclic nucleotide-dependent Na^+/Ca^{2+} exchange serves as a universal sensor through which the biological effect of extremely low concentrations of substances is realized.

As NIR, having less intensity than thermal threshold, dramatically changes the physicochemical properties of water, it is predicted that the cyclic nucleotide-dependent Na^+/Ca^{2+} exchange could also serve as a metabolic gate through which NIR is realized.

Our previous study performed on neuronal and muscle cells has shown that the SMF and ELF EMF-treated PS elevates cGMP-dependent Ca^{2+} efflux from the cell, which is accompanied by the increase of ouabain receptors affinity to α_3 isoform of Na^+/K^+-ATPase in the membrane [10, 46, 47], while the microwaves (MW) have opposite effect on the aforementioned mechanisms. The MW radiation activates cAMP-dependent R Na^+/Ca^{2+} exchange, which is accompanied by the decrease of ouabain receptors affinity to α_3 isoform of Na^+/K^+-ATPase [48, 49].

Thus, on the basis of the above presented data, it can be concluded that water molecule dissociation is a primary target for NIR, while net water efflux through the membrane is a primary signal for the metabolic pathway through which the biological effect of EMF on cells and organisms is realized.

ACKNOWLEDGMENTS

This work was supported by Armenian Governmental Grants, European Office of Aerospace Research and Development Research Grant

Project No#A -803P and European Office of Aerospace Research and Development Research Grant Project No #1592P. We express our gratitude to Ani Gyurjinyan for editing the chapter.

Declaration of Interest

Author has declared that no competing interests exist.

References

[1] Hodgkin, A. L. *The Conduction of the nervous impulse*. 1st edition. Springfield: Liverpool University Press; 1964.

[2] Troshin, A. S. *Problems of Cell Permeability*. 1st edition. Oxford: Pergamon Press; 1966.

[3] Ling, G. N. *In Search of the Physical Basis of Life*. 1st edition. New York: Plenum Publication Co. 1984.

[4] Ling, G. N. (2006). A Convergence of experimental and theoretical breakthroughs affirms the PM theory of dynamically structures cell water on the Theory's 40th Birthday. In G. H. Pollack, I. L. Cameron, D. N. Wheatley (Eds), *Water and the cell* (1st edition, pp. 1-52). Dordrecht, the Netherlands: Springer.

[5] Devyatkov, N. D. (1973). Effect of a SHF (mm-band) radiation on biological objects. *Uspekhi Fizicheskikh Nauk* 110:453-454.

[6] Adey, W. R. (1981). Tissue interactions with non-ionizing electromagnetic field. *Physiol Rev* 61(2):435-514.

[7] Ayrapetyan, S. N. & Carpenter, D. O. (1991). Very low concentration of Acetylcholine and GABA modulate transmitter responses. *Membrane and Cellular Biophysics and Biochemistry NeuroReport* 2(10):563-565.

[8] Azatian, K. V., Ayrapetyan, S. N., Carpenter, D. O. (1997). Metabotropic GABA receptors regulate Acetylcholine responses on snail neurons. *Gen Pharmacol* 29(1):67-72.

[9] Ayrapetyan, S. (2006). Cell aqua medium as a preliminary target for the effect of electromagnetic fields. In S. Ayrapetyan, & M. Markov (Eds.), *Bioelectromagnetics: Current Concepts* (1st edition, pp. 31-64). Dordrecht, the Netherlands, NATO Security through Science Series: Springer Press.

[10] Ayrapetyan, S. N. (2015). The Role of cell hydration in realization of biological effects of non-ionizing radiation (NIR). *Electromagnetic biology and medicine* 34(3):197-210.

[11] Ayrapetyan, S. (2017). The intracellular signaling system controlling cell hydration as a biomarker for EMF dosimetry. In: M. Markov, editor. *Dosimetry in Bioelectromagnetics* (1st edition, pp. 339-366). USA, CRC Press.

[12] Klassen, V. I. *Magnetization of Water Systems*. 1st edition. Moscow: Chemistry Press; 1982.

[13] Borgnia, M., Nielsen, S., Engel, A., Agre, P. (1999). Cellular and molecular biology of the aquaporin water channels. *Annu Rev Biochem* 68:425-58.

[14] Hoffmann, E. K., Sørensen, B. H., Sauter, D. P., Lambert, I. H. (2015). Role of volume-regulated and calcium-activated anion channels in cell volume homeostasis, cancer and drug resistance. *Channels* (Austin) 9(6):380-96.

[15] Ayrapetyan, S. N. & Sulejmanian, M. A. (1979). On the pump-induced cell volume changes. Comp Biochem. *Physiol part A Comp Pharmacol* 64(4):571-575.

[16] Lang, F. (2007). Mechanisms and significance of cell volume regulation. *J Am Coll Nutr* 26:613S–623S.

[17] Lehninger, A. L. *Biochemistry. The molecular basis of cell structure and function*. 1st edition. New York: Worth Publishers, Inc.; 1972.

[18] Ayrapetyan, S. N., Suleymanyan, M. A., Saghyan, A. A., Dadalyan, S.S. (1984). Autoregulation of electrogenic sodium pump. *Cell Mol Neurobiol* 4(4):367- 383.

[19] Ayrapetyan, S. N., Rychkov, G. Y., Suleymanyan, M. A. (1988). Effects of water flow on transmembrane ionic currents in neurons of Helix Pomatia and in squid giant axons. *Comp Biochem Physiol* 89A(2):179-186.

[20] Parsegian, V. A., Rand, R. P., Rau, D. C. (2000). Osmotic stress, crowding, preferential hydration, and binding: A comparison of perspectives. *Proc Nat Sci* USA 97:3987-3992.

[21] Ayrapetyan, S. (2001). Na-K Pump and Na:Ca exchanger as metabolic regulators and Sensors for Extraweak Signals in Neuromembrane. In S.N. Ayrapetyan & A.C.T North (Eds.), *Modern Problems of Cellular and Molecular Biophysics* (1st edition, pp. 31-57). Yerevan, Noyan Tapan.

[22] Ayrapetyan, S. (2012). Cell hydration as a universal marker for detection of environmental pollution. *Environmentalist J.* 32(2):210-221.

[23] Ayrapetyan, S. N., Arvanov, V. L., Maginyan S. N, Azatyan, K. V. (1985). Further study of the correlation between Na-Pump activity and membrane chemosensitivity. *Cell Mol Neurobiol* 5(3):231-243.

[24] Pullman B. & Pullman A. *Quantum Biochemistry.* 1st Edition. New York: Interscience Publisher; 1963.

[25] Baghdasaryan N., Mikayelyan Y., Barseghyan S., Dadasyan E., Ayrapetyan S. (2012). The Density-dependency of dark and low background radiation effects on water and water solution properties. *Electromagnetic biology and medicine* 31(1):87-100.

[26] Chaplin, M. F. (2006). Information exchange within intracellular water. In G. H. Pollack, I. L. Cameron, D. N. Wheatley (Eds.), *Water and the cell* (1st edition, pp. 113-123). Dordrecht, the Netherlands: Springer.

[27] Drost-Hansen, W. (1956). Temperature anomalies and biological temperature optima in the process of evolution. *Naturwissenschaften* 43(22):512-512.

[28] Ayrapetyan, S. N. & Markov M. S. Bioelectromagnetics: *Current Concepts. NATO Security through Science Series, Series B: Physics and Biophysics.* 1st edition. The Netherlands: Springer Press; 2006.

[29] Rychkov, G. Y., Suleymanyan, M. A., Ayrapetyan, S. N. (1989). Dependence of water flow effect on the ionic currents of dialyzed neuron of somatic membrane fluidity. *Biol Membr* 6(7):733-739.

[30] Suleymanian, M. A., Ayrapetyan, V. Y., Arakelyan, V. B., Ayrapetyan, V. Y. (1993). The effect of osmotic gradient on the outward potassium current in dialyzed neurons of Helix Pomatia. *Cell Mol Neurobiol* 13(2):183-190.

[31] Ayrapetyan, S. N. & De, J. (2014). Cell hydration as a biomarker for estimation of biological effects of non-ionizing radiation on cells and organisms. *The Scientific World Journal* 2014(890518):1-8.

[32] Ayrapetyan, S. N., Rychkov, G. Y., Suleymanian, M. A. (1988). Modulation of calcium currents in the snail neuron membrane by osmotic gradient. *Dokl Akad Nauk SSSR* 300(4):983–985.

[33] Carpenter, D. O., Fejtl, M., Ayrapetyan, S. N., Szarowski, D. H., Turner, J. N. (1992). Dynamic changes in neuronal volume resulting from osmotic and sodium transport manipulations. *Acta Biologica Hungarica* 43(1-4):39-48.

[34] Sagian, A. A., Ayrapetyan, S. N., Carpenter, D. O. (1996). Low concentrations of ouabain stimulate Na:Ca exchange in neurons. *Cell Mol Neurobiol* 16(4):489-98.

[35] Juhaszova, M. & Blaustein, M. P. (1997). Na+ pump low and high ouabain affinity a subunit isoforms are differently distributed in cells. *Proc Natl Acad Sci* USA 94:1800-1805.

[36] Blaustein, M. P. & Lederer, W. J. (1999). Na+/Ca2+ exchange. Its physiological implications. *Physiol Rev* 79:763-854.

[37] Xie Z., Askari, A. (2002). Na+/K+-ATPase as a signal transducer. *Eur J Biochem* 269:2434–2439.

[38] Blackman, C. F., Elder, J. A., Weil, C. M., Benane, S. G, Eichinger, D. C., House, D. E. (1979). Induction of calcium ion efflux from brain tissue by radiofrequency radiation: Effects of modulation-frequency and field strength. *Radio Science* 14(6S):93-98.

[39] Azatian, K. V., White, A. R., Walker, R. J., Ayrapetyan, S. N. (1998). Cellular and molecular mechanisms of nitric oxide-induced heart muscle relaxation. *Gen Pharmac* 30(4):543-553.

[40] Brini, M. & Carafoli, E. (2009). Calcium pumps in health and disease. *Physiol Rev* 89(4):1341-1378.

[41] Arvanov, V. L., Tsai, M. C., Walker, R. J., Ayrapetian, S. N. (1993). Interaction of concanavalin A and wheat germ Agglutinin with Helix Acetylcholine receptors. *Brain Res* 615(2):252-258.

[42] Suleymanian, M. A., Takenaka, T., Stamboltsyan, K. V., Ayrapetyan, S. N. (1986). The effects of short-chain fatty acids on the neuronal membrane functions of Helix Pomatia. *I Electrical Properties Cell. and Moll. Neurobiol*. 6:(2)151-163.

[43] Takenaka, T., Horie, H., Kawasaki Y. (1983). Effect of fatty acids on the membrane fluidity of cultured chick dorsal root ganglion measured by fluorescence photo bleaching recovery. *J Neurobiol* 14:457-461.

[44] Azatian, K. V., Karapetian, I. C., Ayrapetyan, S. N. (1993). Effect of Acetylcholine of low doses on 45Ca influx into the Helix Pomatia neurons. *Biological Membranes* 10(3):317-320.

[45] Grigorian, K. P., Azatyan, K. V., Kazaryan, S. H., Ayrapetyan, S. N. (2001). Ouabain-sensitive and insensitive Acetylcholine receptors on the membrane of the same neuron in Helix Pomatia. *Comparative Biochemistry and Physiology C Toxicol Pharmacol* 128(4):511-20.

[46] Ayrapetyan, S. N., Avanesian, A. S., Avetisian, T. H., Majinian, S. B. (1994). Physiological effects of magnetic fields may be mediated through actions on the state of calcium ions in solution. In D. O. Carpenter, & S. N. Ayrapetyan (Eds.), *Biological Effects of Electric and Magnetic Fields: Sources and mechanisms* (1st edition, Vol. 1, pp. 181-192). California: Academic Press.

[47] Ayrapetyan, G., Papanyan, A., Hayrapetyan, H., Ayrapetyan, S. (2005) Metabolic pathway of magnetized fluid-induced relaxation effects on heart muscle. *Bioelectromagnetics* 26(8):624-630.

[48] Nikoghosyan, A., Heqimyan, A., Ayrapetyan, S. (2016). Non-thermal microwave radiation-induced brain tissue dehydration as a potential factor for brain functional impairment. *Int J Basic App Sci* 5(4):188-195.

[49] Narinyan, L. & Ayrapetyan, S. (2017). Cyclic AMP-dependent signaling system is a primary metabolic target for non-thermal effect of microwaves on heart muscle hydration. *Electromagnet Biol Med* 36(2):182-191.

BIOGRAPHICAL SKETCH

Prof. Sinerik N. Ayrapetyan, PhD

Place of Birth: Armenia

Nationality: Armenian

Business address: UNESCO Chair in Life Sciences, Life Sciences International Postgraduate Educational Centre (LSIPEC), Avan – 31 Acharian St., Yerevan 0040, Armenia

Position Held: Head of Research Council of LSIPEC and Coordinator of UNESCO Chair
E-mail: life@biophys.am info@biophys.am

Education:

Degrees	Dates	University
Master of Sciences	1960 -1966	Yerevan State University
PhD	1966 – 1969	Bogomolets Inst. Physiology, Acad. Sci. Ukraine.
Doctor of Sciences	1970 – 1980	Bogomolets Inst. Physiology, Acad. Sci. Ukraine.

Professional Appointments:

Employers	Position	Dates
Life Sciences International Postgraduate Educational Center	Chairholder of UNESCO Chair in Life Sciences	1997-present

Life Sciences International Postgraduate Educational Center	Head of Research Council	2010-present
Life Sciences International Postgraduate Educational Center	President	1997 - 2010
Armenian National Academy of Sciences	Head of the Biophysics Center	1989 - 1997
Institute of Experimental Biology, Arm. NAS	Head of Membranology Laboratory	1973 - 1989
Department of Physiology, Yerevan State University	Associate Professor	1971 - 1973
The Institute of Physiology of the Armenian Academy of Sciences	Senior Researcher	1969 - 1971
Visiting Researcher		
Department of Pharmacology and Physiology, Southampton University	Visiting Professor	1996 (a month)
Department of Life Science, Nottingham University, UK	Visiting Professor	1995 (3 months)
SUNY at Albany, New York	Visiting Professor	1989-1992
National Institute of Physiological Sciences, Okazaki, Japan	Visiting Researcher	1981 (6 months)
Physiological Dept. of King's Collage, London, UK	Visiting Researcher	1980 (3 months)
Physiological Department of Belgrade University, Yugoslavia	Visiting Researcher	1980 (3 months)
Institute of Biology, Hungarian Academy of Sciences, Tihany	Visiting Researcher	1975 (3 months)

Specialization:

Main field: biophysics, neuroscience

Other fields: ionizing and non-ionizing radiobiology

Current research interest: intracellular signaling system in norm and pathology

Membership of Professional Societies

Member of: International Society of Invertebrate Neurobiology (ISIN)
European Society for Neurochemistry (ESN)

International Society for Neurochemistry (ISN)

International Brain Research Organization (IBRO)

International Union of Pure and Applied Biophysics (IUPAB)
Bioelectromagnetics Society (BEMS)

WHO International Advisory Committee on Electromagnetobiology

President of: All Armenian Research Council

Editor in Chief of Journals:

Bioavailability and Bioequivalence
Biomedical Engineering Current Research
Basic, Applied Pharmacy and Pharmacology

Member of Editorial Board of:

Journal of Electromagnetic Biology and Medicine
Journal of ISRN Biophysics
Journal of European Journal of Biophysics
Journal of Advances in Life Sciences
Journal of Applied Pharmacy
Journal of BBA General Subjects
Journal of International Dental and Medical Research
Journal of International Journal of Basic and Applied Sciences
Journal of Insights of Medical Sciences
Journal of Chronicles of Pharmaceutical Science Journal
Journal of Clinical Investigation
Journal of Global Drugs and Therapeutics

Journal of Genetic Engineering and Biotechnology
Journal of Basic, Applied Pharmacy and Pharmacology
Journal of Targeted Drug Delivery
Journal of Frontiers of Mechatronical Engineering

Selective Publications:

Books

1) *Bioelectromagnetics: Current Concepts* 2006 (Eds. S. Ayrapetyan (Armenia) & M. Markov (USA) NATO Science Series, Springer Press. 480 pages.
2) *Agricultural Biotechnology and Biosafety for Food Security in the Caucasus Region and Moldova* 2004 (Eds. K. Nichterlein (FAO, Italy) & S.N. Ayrapetyan (Armenia) Noyan Tapan Press, 175 pages.
3) *Modern Problems of Cellular and Molecular Biophysics* 2001 (Eds. Ayrapetyan S. N. (Armenia) & North A.C. T. (IUPAB, UK), 360 pages.
4) *Pain Mechanisms and Management* 1998 (Eds. Ayrapetyan S.N. (Armenia) & Apkarian A.V. (USA), IOS press, Netherlands, 390 pages.
5) *Biological Effects of Electric and Magnetic Fields – Sources and Mechanisms volume 1*, 1994 (Eds. D. O. Carpenter (USA) & S.N. Ayrapetyan (Armenia), Academic Press, New York.
6) *Biological Effects of Electric and Magnetic Fields – Beneficial and Harmful Effects volume 2*, 1994 (Eds. D. O. Carpenter (USA) & S.N. Ayrapetyan (Armenia), Academic press, New York.
7) *Metabolic Regulation of Membrane Function* 1990 (Eds. Ayrapetyan S. N Arvanov V. L, Arutchyan N, I, Kalantaryan L.B., Suleymanyan M.A), Armenian Academy of Sciences Publishers, Yerevan.
8) *The Membrane Ionic Transport upon the Effect of Ionic Radiation* 1990, Dvoretsky, A. I, Ayrapetyan, S. N., Shainskaya, A. M., Chebotarev, Y. Y., Naukovo Dumka, Kiev.

Research Articles

2017

Ayrapetyan S. (2017) (Editorial) The cyclic nucleotides-dependent Na^+/Ca^{2+} exchange-generated water efflux from the cell as a primary metabolic target for pain therapy, *Chronicles of Pharmaceutical Science*, 1(6):356-359.

Ayrapetyan S., Heqimyan A., Nikoghosyan A. (2017) The comparative study of 8Hz EMF effect on tissue hydration in brain cortex and subcortex of rats. *Advances in Life Sciences* 7(3): 31-38.

Ayrapetyan S. (2017) The intracellular signaling system controlling cell hydration as a biomarker for EMF dosimetry, In: *Dosimetry in Bioelectromagnetics*, ed. M. Markov, CRC Press, USA pp. 339-366.

Ayrapetyan S. (2017) Reverse mode Na^+/Ca^{2+} exchange-induced cell dehydration as a primary mechanism for cell pathology. *Global Drugs and Therapeutics* 2(6): 1-4.

Ayrapetyan S. (2017) Lipid fluidity as an essential therapeutic tool for cell pathology. *Bioequiv Availab*, 9:3 10000e80, DOI: 10.4172/ jbb.10000e80.

Narinyan L., Ayrapetyan S. (2017) Cyclic AMP-dependent signaling system is a primary metabolic target for non-thermal effect of microwaves on heart muscle hydration, *Electromagnetic Biology and Medicine*, 36(2):182-191; DOI: 10.1080/15368378.2016.1241803.

Musheghyan G., Minasyan A., Arajyan G., Ayrapetyan S. (2017) 4Hz Mechanical vibration relieves pain through Na^+/K^+ATP-ase α_3 isoform-dependent brain tissue dehydration. *International Journal of Basic and Applied Sciences* 6 (2):29-35.

2016

Nikoghosyan A., Heqimyan A., Ayrapetyan S. (2016) Non-thermal microwave radiation-induced brain tissue dehydration as a potential factor for brain functional impairment, *International Journal of Basic and Applied Sciences*, 5(4):188-195 doi: 10.14419/ijbas.v5i4.6430.

Ayrapetyan S. (2016) *Cell hydration-induced changes of membrane conductivity as a marker for estimation of biological effects of chemical and physical factors on organism*, 8:e72. doi:10.4172/jbb.10000e72.

Ayrapetyan G, Hayrapetyan H, Ayrapetyan S. (2016) The Effect of non-thermal microwave-treated physiological solution on isolated heart muscle of snail, *Current Chemical Biology*, 10(1):2-8.

Ayrapetyan S. (2016) (Short Communication) The nM ouabain-induced tissue dehydration as a novel diagnostic marker for neuronal pathology, *Global Drugs and Therapeutics*, 2(1):1-2. doi: 10.15761/GDT.1000108.

2015

Ayrapetyan S. (2015) The Dysfunction of cAMP-dependent Na^+/Ca^{2+} Exchange in Reverse Mode as a Primary Mechanism for Age-dependent Cardio-muscle Failure, *J Bioequiv Availab*, 8:1, http://dx.doi.org/10.4172/jbb.10000e70.

Ayrapetyan S. (2015) The Dysfunction of Metabolic Controlling of Cell Hydration Precedes Warburg Phenomenon in Carcinogenesis. *J Bioequiv Availab* 7:e59. doi: 10.4172/jbb.10000e59.

Ayrapetyan S.N (2015) The Role of Cell Hydration in Realization of Biological Effects of Non-ionizing Radiation (NIR). *Electromagnetic biology and medicine*. 34(3):197-210.

Narinyan L. Ayrapetyan S. (2015) Dysfunction of nM Ouabain-Induced Activation of the Signaling System Responsible for Age-Related Heart Muscle Failure. *Advances in Life Science* 5(4):73-84 DOI: 10.5923/j.als.20150504.01.

Nikoghosyan A., Heqimyan A., Ayrapetyan S. (2015) Primary Mechanism Reponsible for Age-Dependent Neuronal Dehydration. *International Journal of Basic and Applied Sciences*, 5(1):5-14. doi: 10.14419/ijbas.v5i1.5388.

In: Ionizing Radiation
Editor: Tamar Reeve
ISBN: 978-1-53613-539-8

Chapter 7

RADIOPROTECTIVE PROPERTIES OF ASTAXANTHIN: THE IMPACT ON RADIATION INDUCED CHROMOSOMAL ABERRATIONS AND DNA BREAKS IN HUMAN LYMPHOCYTES *IN VITRO*

Stanislav R. Rushkovsky[1,*], Denis A. Kurinnyi[2], Olena M. Demchenko[2] and Maria A. Pilinska[2]

[1]Department of General and Medical Genetics, Educational and Scientific Centre "Institute of Biology and Medicine," Taras Shevchenko National University of Kyiv, Kyiv, Ukraine

[2]Department of Medical Genetics, State Institution "National Research Center for Radiation Medicine of the National Academy of Medical Sciences of Ukraine," Kyiv, Ukraine

---

[*]Corresponding Author Email: rsr@ukr.net.

Abstract

*Purpose:*Astaxanthin (a carotenoid of xanthophyll group) satisfies all requirements demanded to potential radioprotectors. The purpose of our study was to investigate a modifying effect of astaxanthin on radiation induced genome damages in cultivating human peripheral blood lymphocytes (PBL).

Materials and methods: To establish astaxanthin radioprotective activity, we used method for scoring of structural chromosomal aberrations (ChA) in PBL and neutral single-cell gel electrophoresis (comet assay) for estimation of the level of DNA breaks in cells (expressed as tail moment (TM) for "classical comets" and rate of "atypical comets" (AC) for cells with greatly fragmented DNA). Blood samples were obtained by venepuncture from five conditionally healthy volunteers. Astaxanthin in final concentrations of 2.0, 10.0, 20.0 and 40.0 μg/ml had been added into the culture medium before cultivation began, prior to γ- irradiation of PBL (dose 1.0 Gy). Cultures were incubated at 37°C for 48-50 h. Chromosome slides were prepared after 50 h of cultivation. Comet assay was carried out before cultivation and on 48 h of culturing.

*Results:*Astaxanthin in all studied concentrations *per se* had no mutagenic activity and obviously reduced the level of radiation induced chromosomal aberrations. The most efficient concentration of astaxanthin was 20.0 μg/ml. The effect of astaxanthin in this concentration on exposed cells manifested in significant decreasing of the radiation-induced ChA level: from 26.06 ± 1.81 per 100 metaphases (phm) to 9.03 ± 0.73 phm (p<0,001). Comparative analysis of the ChA spectra indicated the most substantial reduction of classic unstable cytogenetic markers of radiation exposure (dicentric and ring chromosomes). Likewise, statistically significant decrease in TM in irradiated lymphocytes treated by astaxanthin compared with untreated irradiated samples was observed both before cultivation (3.21 ± 0.48 and 6.55 ± 1.82 respectively, p<0,05) and after 48 h of cultivation (5.27 ± 1.77 and 12.86 ± 0.74 respectively, p<0,05). The level of AC, which evidently arose from apoptotic cells, in irradiated cultures treated by astaxanthin was approximately two times higher as compared with irradiated samples (7.15 ± 1.13% and 3.57 ± 0,81% respectively, p<0,05).

*Conclusions:*Our study indicated that astaxanthin in concentration of 20.0 μg/ml demonstrated evident radioprotective properties by reducing the level of ChA, decreasing DNA damage and increasing the apoptotic rate.

Keywords astaxanthin, human peripheral blood lymphocytes, chromosomal aberrations, DNA breaks, apoptosis, radioprotective effect

INTRODUCTION

The accidents at the nuclear power plants in Chernobyl and Fukushima, as well as increased in recent years the threat of nuclear terrorism promote intensive investigations to search effective and safe bioactive substances, preferably of natural origin, that may be used as radiation protectors. Some carotenoids, for example β-carotene and lycopene, have a radioprotective activity concerning to the mutagenic and carcinogenic effects of ionizing radiation (Srinivasan et al. 2009, Berti et al. 2014).

Astaxanthin is a carotenoid of xanthophyll group, and it is one of the most common red pigments of algae, yeasts, krill, shrimps, crayfish, trout and salmon (Ambati et al. 2014). Astaxanthin is a powerful antioxidant that scavenges and quenches reactive oxygen species in 550 times more effectively than α-tocopherol and in 16 times than β-carotene (Nishida et al. 2007). According to large number of studies, astaxanthin is ableto reduce oxidative DNA damages *in vitro* and *in vivo* (Azqueta and Collins 2012), to inhibit UV- and chemically induced neoplastic cell transformation (Rao et al. 2013) and cancer cell proliferation (Palozza et al. 2009, Li et al. 2015, Kim et al. 2016), to stimulate apoptotic processes in cancer cells (Song et al. 2011, Li et al. 2015), to exert positive influence on immune system functioning (Lin et al. 2015) and to manifest anti-inflammatory activity (Kim et al. 2016). Moreover, as opposed to β-carotene, astaxanthin is not a vitamin precursor (Krinsky 1994), and evidently no toxicity of high doses of astaxanthin has been demonstrated in studies using laboratory animal models (Tago et al. 2014).

These properties of astaxanthin suggest that it may possess antimutagenic activity and thus it satisfies all requirements demanded to potential radioprotectors. In spite of numerous studies of astaxanthin beneficial actions there is no evidence of its possible geno- and radioprotective properties.

The purpose of our study was to investigate a modifying effect of astaxanthin on radiation induced genome damages in cultivating human peripheral blood lymphocytes (PBL). For evaluation of astaxanthinradioprotective activity we used well established method for scoring of structural chromosomal aberrations (Tawn and Holdsworth 1986) and neutral single-cell gel electrophoresis (comet assay), as a widely accepted approach to estimate the level of single-strand and double-strand breaks in cells (Olive and Banáth 2006).

Structural chromosomal aberrations arise as a result of DNA double-strand breaks or other kinds of DNA damage which may be easily converted in double-strand breaks (e.g., single-strand breaks or DNA cross-links). Insufficient repair of double strand breaks leads to deletions or aberrant chromosomal exchanges (Natarajan and Palitti 2008, Obe and Durante 2010). The comet assay is a widely used method to evaluate the level of DNA damage in cells (Olive and Banáth 2006, Liao et al. 2009). There are two main variants of this method: the neutral comet assay (performed at neutral pH) and the alkaline comet assay (performed at pH>13). It is accepted that the neutral comet assay detects double-strand and single-strand breaks in DNA (causes relaxation of supercoiled loop domains and facilitates comet tail formation) and double-strand DNA fragments (Afanasieva et al. 2010, Afanasieva et al. 2013). DNA single-strand and double-strand breaks are the main types of DNA damage induced by ionizing radiation. Some of them originate immediately in the course of radiation exposure, another part of DNA breaks may arise within the lapse of time after irradiation by conversion from the rest of DNA damage during their reparation, replication or transcription (Little 2003, Aguilera and García-Muse 2013). Occurrence of unrepaired or misrepaired DNA breaks (“immediate” or “delayed”) results in chromosomal aberrations (as was mentioned above), cell cycle arrest and cell death (Heijink 2013). On these accounts, for proper comparison of data obtained by comet assay with results of cytogenetic studies, we preferred neutral variant of the comet assay and chose two points in time for DNA damage detection: 0 hours and 48 hours (the same time for chromosome harvesting) after irradiation of blood samples. In addition, neutral variant of the comet assay allows us to exclude cells that are in S-phase of cell cycle from the pool of damaged cells (Olive and Durand 2005) which may occur after 48 hours of culturing.

MATERIAL AND METHODS

Blood Collection

Blood samples were obtained by venepuncture from five conditionally healthy volunteers (3 men and 2 women) at the age of 20-51 years old, average age was 41 years. Only the persons, who denied any conscious contact with well-known or potential mutagenic agents, were not hard smokers or hard drinkers and had not any inherent or acquired pathology were chosen for the investigation. All persons were involved in research under conditions of informed consent.

Cultivation of Human Lymphocytes and Preparation of Chromosome Slides

Samples of whole blood (0.5 ml) were cultivated in 4.5 ml of RPMI-1640 medium with 10 μg per ml of PHA-P (Sigma, USA). Cultures were incubated at 37°C for 50 h. To estimate radioprotective properties of astaxanthin some cultures were exposed to γ-ray (dose 1 Gy, emitter IBL-237C, dose-rate 2.34 Gy/min) on first hour of cultivation. Astaxanthin (Sigma, USA) was added into culture medium in final concentrations of 2.0 μg/ml, 10.0 μg/ml, 20.0 μg/ml and 40.0 μg/ml before gamma irradiation of lymphocytes. The concentrations of astaxanthin were chosen based on literature data (Azqueta and Collins 2012) and results of our preliminary experiments. Prior to the end of the culturing (1.5 - 2 hours), colchicine (1 µg/ml, final concentration) was added to each culture to arrest mitosis. After cultivation, cells were treated with 0.075 M KCl solution and fixed in four changes of methanol:glacial acetic acid mixture (3:1). To prepare metaphase plates, the cell suspension was dropped onto cool moist slides and dried under room temperature.

Chromosome Analysis

All slides were uniformly stained with 2% Giemsa mixture in distilled water for 20 min. (without banding). Well-spread metaphases (more than 150 for each volunteer/variant) were scored under the light microscope (×1000 magnification) for chromatid and chromosome type lesions. Scoring of chromatid breaks was performed according to CBIC recommendation, chromatid gaps were not scored according to generally accepted criteria (Tawn and Holdsworth 1991, Hsu et al. 1996). The frequency of chromosomal aberrations per 100 metaphases (phm) was chosen as a main parameter of genome damage. Metaphase plates with less than 44 chromosomes were not considered.

Single-Cell Gel Electrophoresis (Comet Assay)

For comet assay we used lymphocytes from whole fresh blood samples (without culturing) and from 48-hour human PBL cultures. The blood samples and PBL cultures were treated similarly as described above. The lymphocytes were isolated by separation of cells in a density gradient Histopaque 1077 (Sigma, USA), according to instructions of the manufacturer, and then washed in 0.9% NaCl twice. 50 µl of the suspension was mixed with 100 µl of 1% low-melting point agarose (Sigma, USA) at 37°C. 60 µl of the mixture was used to prepare a microscope slide, covered with 1% high-melting point agarose (Sigma, USA). After agarose polymerization at 4°C, slides were immersed for several hours in ice-cold lysis solution (2.5 M NaCl, 100 mM EDTA, 10 mMTris–HCl (pH 8.0), and 1% Triton X-100 (Ferak, Germany), which was added just before use) for several hours. After the lysis slides were washed by TBE buffer (89 mMTris–borate, 2 mM EDTA, pH 7.5).

Slides were electrophoresed (1 V/cm, 200 mA, 15 min.) at 4°C in TBE buffer. After electrophoresis, slides were stained with 2 μg/ml of DAPI (4′,6-diamidino-2-phenylindole, Sigma, USA) and immediately analyzed with a fluorescent microscope Opton Axioskop (Zeiss, Germany), connected with a Canon EOS 1000 D camera. A total of 100–200 randomly chosen cells with the exception of "atypical comets" (AC) (Figure 1) on each slide were examined using image processing program ImageJ (imagej.nih.gov) with OpenComet plugin (Gyori et al. 2014). Tail moment (TM) was chosen as a parameter to determine the relative level of DNA damage. The AC, which are apparent apoptotic cells (Roser et al. 2001), were scored separately.

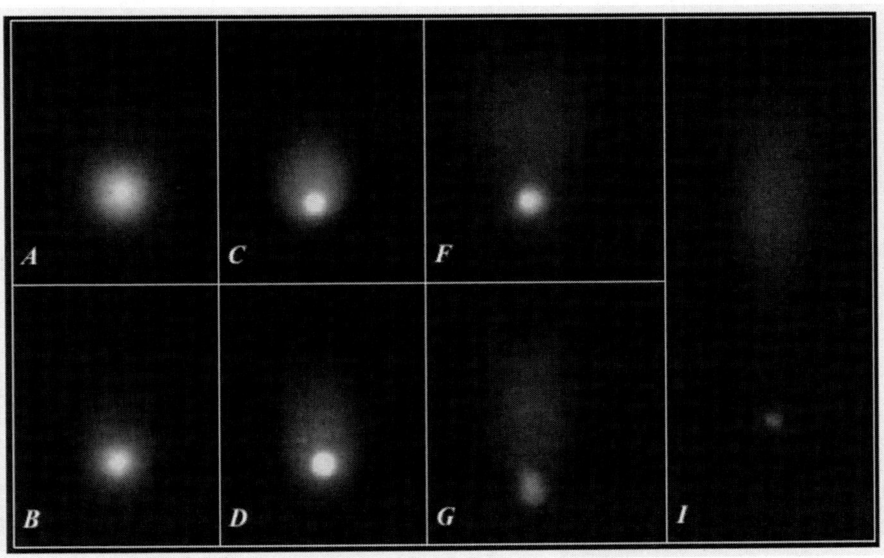

Figure 1. Examples of comets obtained from cells with different amount of DNA breaks (the comet on I represents "atypical comet").

Statistical Analysis

Statistical processing was carried out by the standard methods both for cytogenetic analysis and comet assay (Rosner 2015). Significance of differences between groups by levels of chromosomal aberrations was estimated by the 2×2 chi-square (χ^2) test. Data from the comet assay was analyzed by one-way ANOVA using QtiPlot software (version 0.9.8.8). A p value less than 0.05 was considered statistically significant.

RESULTS

The Influence of Astaxanthin on Spontaneous and γ -Radiation Induced Chromosomal Aberrations

The spontaneous frequency of chromosomal aberration (ChA) in studied group varied from 1.33 to 3.02 *per* 100 metaphases (phm) with average level of 2.47 ± 0.30 phm. The main chromosomal abnormalities were single and double acentric fragments (1.48 ± 0.27 phm and 0.97 ± 0.17 phm, respectively). Supplementation with astaxanthin did not cause significant changes in levels of ChA in cultures of nonirradiated lymphocytes (Table 1).

After radiation exposure *in vitro* we have found significant increase of ChA in lymphocytes of all donors in comparison with spontaneous levels: the mean frequency of ChA was 26.06 ± 1.81 phm ($p<0.001$) with individual variations from 22.11 phm to 32.67 phm. The aberrations of chromosome type were prevalent among chromosome damage (Figure 2), which is typical for chromosomal mutagenesis after exposure to ionizing radiation in G0/G1 phase of the cell cycle: the levels of chromosome type ChA and chromatid type ChA were 24.45 ± 1.69 phm and 1.61 ± 0.30 phm respectively.

Table 1. The influence of astaxanthin on spontaneous and γ-radiation induced frequencies of chromosomal aberrations in human peripheral blood lymphocytes *in vitro*

Concentrations of astaxanthin, μg/ml	Chromosomal aberrations, phm (X ± Se)	
	Nonirradiated cultures	Irradiated cultures
0	2.47 ± 0.30	26.06 ± 1.81
2.0	2.61 ± 0.17	18.00 ± 2.72*
10.0	2.41 ± 0.26	13.67 ± 1.98**
20.0	2.34 ± 0.20	9.03 ± 0.73***
40.0	2.56 ± 0.19	12.50 ± 2.34**

Notes: * - $p<0.05$, ** - $p<0.01$, *** - $p<0.001$ compared with irradiated samples without astaxanthin treatment.

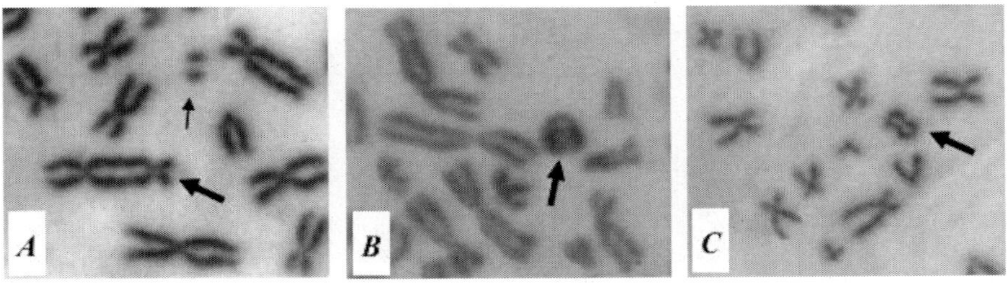

Figure 2. Examples of main chromosomal aberrations induced by γ-radiation:
A - dicentric chromosome (thick arrow) with acentric double fragment (thin arrow);
B - centric ring chromosome (arrowed); *C* - acentric rings (arrowed).

Effect of astaxanthin on exposed cells manifested in significant decrease of the average level of radiation induced ChA for all chosen concentrations. The most efficient final concentration was 20.0 μg/ml (Table 1): the mean frequency of aberrations was reduced approximately 3 times, up to 9.03 ± 0.73 phm ($p<0.001$ compared to irradiated samples without astaxanthin treatment). The reduction of ChA was observed in cell cultures of all volunteers and varied from 7.69 phm to 11.75 phm. Similar to results obtained for only irradiated samples, the chromosome type aberrations were prevalent in this experiment too (8.41 ± 0.79 phm for chromosome type ChA and 0.62 ± 0.21 for chromatid type ChA) (Table 2).

Table 2. The impact of astaxanthin in concentration of 20.0 μg/ml on spontaneous and γ-radiation induced levels and spectra of chromosomal aberrations in human PBL *in vitro*

Treatment	Chromosomal aberrations, phm (X ± Se)	Chromatid-type aberrations, phm (X ± Se)	Chromosome-type aberrations, phm (X ± Se)				
		Acentric single fragments	Acentric double fragments	Dicentrics	Centric rings	Acentric rings	Abnormal monocentrics
C	2.47 ± 0.30	1.48 ± 0.27	0.97 ± 0.17	0	0	0	0.02 ± 0.02
A	2.34 ± 0.20	1,46 ± 0.30	0.88 ± 0.24	0	0	0	0
Irr	26.06 ± 1.81	1.61 ± 0.30	8.07 ± 1.64	12.26 ± 0.66	2.42 ± 0.44	0.63 ± 0.27	1.06 ± 0.61
Irr+A	9.03 ± 0.73	0.62 ± 0.21	4.34 ± 0.63	3.26 ± 0.78	0.34 ± 0.18	0.16 ± 0.11	0.31 ± 0.16

Notes: C - control (non supplemented with astaxanthin and non irradiated cultures); A - supplemented with astaxanthin cultures; Irr - irradiated cultures; Irr+A - irradiated and supplemented with astaxanthin cultures; X ± Se - mean and standard error.

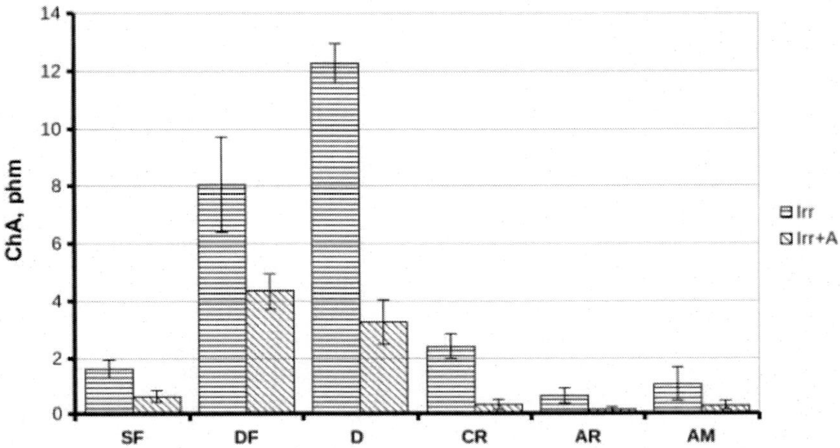

Figure 3. Spectra of chromosomal aberrations in irradiated cultures (Irr) and irradiated cultures supplemented with 20.0 μg/ml astaxanthin (Irr+A). SF – acentric single fragments, DF – acentric double fragments, D – dicentric chromosomes, CR – centric rings, AR – acentric rings, AM – abnormal monocentrics.

Comparative analysis of the ChA spectra indicated the most substantial decrease in the levels of dicentrics and centric ring chromosomes, which are classic unstable cytogenetic markers of radiation exposure (Figure 3). The level of dicentrics was reduced from 12.26 ± 0.68 phm to 3.26 ± 0.78 phm ($p < 0.001$), and the frequency of centric rings was decreased from 2.42 ± 0.44 phm to 0.34 ± 0.18 phm ($p < 0.05$). Statistically significant decrease in double fragments frequency was also found (from 8.07 ± 1.64 phm to 4.34 ± 0.63 phm, $p < 0.05$). Thus, in our subsequent work astaxanthin was used in concentration of 20.0 μg/ml.

The Influence of Astaxanthin on Relative Level of DNA Damages in Human Lymphocytes

The data for mean relative levels of DNA damage (expressed as tail moment - TM) in human lymphocytes before (0 hour) and after cultivation (48-th hour) are presented on the Figure 4.

Before cultivation no increase in DNA damage was detected in nonirradiated samples supplemented with astaxanthin in concentration of 20.0 μg/ml, compared with untreated lymphocytes (TMs were 3.55 ± 1.37 and 2.80 ± 0.57 respectively, $p>0.05$). As expected, significant increasing of TM was detected after γ-irradiation (6.55 ± 1.82, $p<0.05$ compared with control samples). The treatment of cells with astaxanthin prior γ-irradiation resulted in statistically significant decrease in radiation-induced DNA damage (TM=3.21 ± 0.48, $p<0.05$ compared with irradiated samples).

Similar results were obtained after 48 hours of cultivation: statistically significant decrease of TM in irradiated lymphocytes treated by astaxanthin compared with untreated irradiated samples was revealed (TMs were 5.27 ± 1.77 and 12.86 ± 0.74 respectively, $p<0.05$). Also, we observed moderate but insignificant increase of TM in lymphocytes from nonirradiated cultures with astaxanthin compared with control nonirradiated and untreated cells ($p>0.05$).

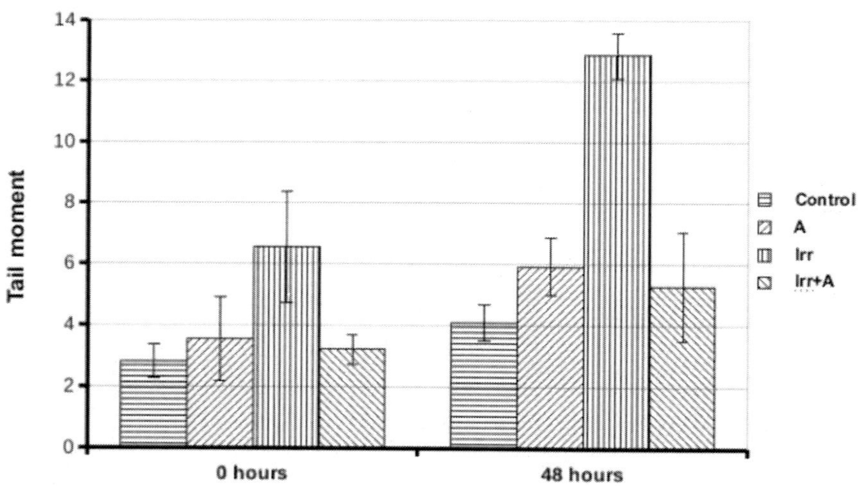

Figure 4. The relative levels of DNA damages in human lymphocytes. C –control (nonsupplemented with astaxanthin and nonirradiated cultures); A - supplemented with 20.0 μg/ml astaxanthin cultures; Irr -irradiated cultures; Irr+A - irradiated and supplemented with 20.0 μg/ml astaxanthin cultures.

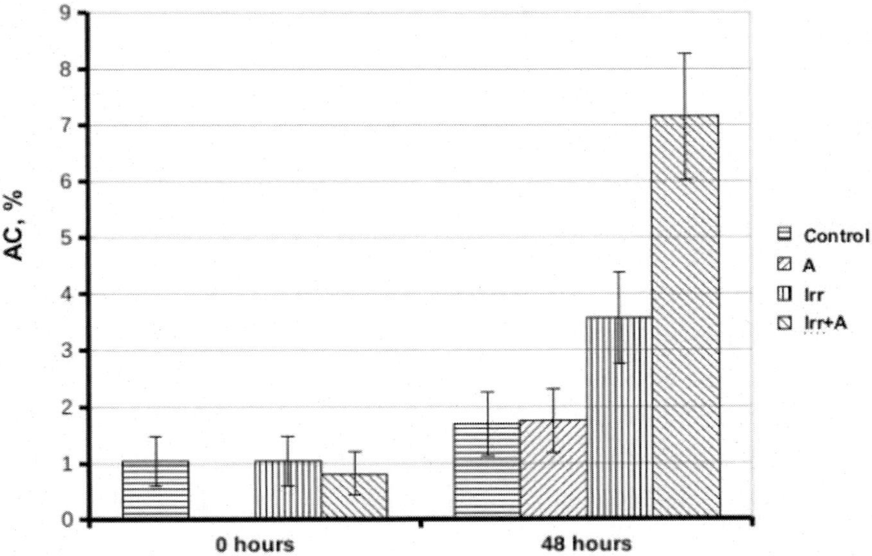

Figure 5. The levels of "atypical comets" (AC) in cultures of human lymphocytes. C - control (nonsupplemented with astaxanthin and nonirradiated cultures); A - supplemented with 20.0 μg/ml astaxanthin cultures; Irr - irradiated cultures; Irr+A - irradiated and supplemented with 20.0 μg/ml astaxanthin cultures.

Because of some limitations of comet analyzing software we used (we found OpenComet recognizes and analyzes "atypical comets" mostly incorrect), the AC were scored separately and their level was chosen as accessory parameter for DNA damage estimation. Before cultivation the AC level was low and did not exceed 1.03 ± 0.44% (Figure 5). There were no statistically significant differences among groups. After cultivation the AC levels have been significantly increased in irradiated samples, both untreated and treated by astaxanthin. Moreover, contrary to TM data, the AC level in irradiated treated with astaxanthin cultures was approximately 4 times higher as compared with 48 hour control (7.15 ± 1.13% and 1.69 ± 0.56% respectively, $p<0.01$) and 2 times higher as compared with irradiated samples (7.15 ± 1.13% and 3.57 ± 0,81% respectively, $p<0.05$).

Discussion

In this study we examined the radioprotective properties of astaxanthin. For this purpose we used two of the most powerful and sensitive methods for detection of radiation induced genome damage both *in vitro* and *in vivo*: the method based on scoring and analysing structural chromosomal aberrations in cultured human blood lymphocytes and the method of single-cell gel electrophoresis (comet assay).

The first step for testing any potential genoprotecors is an estimation of their ability to induce genome damages. Our data indicated that astaxanthin in chosen concentration *per se* doesn't have any mutagenic activity as far as it did not increase the frequencies of chromosomal aberrations and relative levels of DNA damage.

Radioprotective properties of astaxanthin were clearly demonstrated both in cytogenetic studies and by comet assay: abrupt decrease of ChA frequencies in irradiated cultures supplemented with astaxanthin correspond with reduction of radiation-induced levels of TM detected immediately and after 48 hours in the same samples. Inverse data was obtained concerning the frequencies of "atypical comets" (D. A. Kurinnyi et al. 2018). On 48-th hour AC rate was significantly higher in irradiated and treated with astaxanthin cultures than in irradiated only samples. This apparent discordance may be easily explained if we assume that "atypical comets" arise from apoptotic cells (Roser et al. 2001) and the level of AC reflects the apoptotic rate in the cell culture. It is obvious the "atypical comets" form from cells with greatly fragmented DNA. The radiation dose we applied is quite low and cannot induce DNA fragmentation like this (according to Ahnström and Erixon (Ahnström and Erixon 2001) the γ-radiation in dose of 1.0 Gy induces approximately 1000 single-strand and 4 double-strand breaks per cell) that is confirmed by negligible AC levels detected before cultivation irrespective of treatment. So, the "atypical comets" that were observed after 48 hours of cultivation are evidently nuclei of apoptotic cells. It is well known that DNA damage induced by ionizing radiation may lead to apoptosis, and cells bearing unstable chromosomal aberrations are eliminated in the first place (Belloni et al. 2005). Indeed, by analyzing the ChA spectra (Figure 3) we found that supplementation with astaxanthin reduces exactly the levels of classic unstable cytogenetic markers of radiation exposure. Thus, radioprotective properties of astaxanthin showed themselves by reducing the ChA level, decreasing the level of DNA damage and increasing the apoptotic rate.

Comparison of data obtained by cytogenetic analysis and comet assay allows us not only to conclude about radioprotective properties of astaxanthin, but also to make suggestion about mechanisms of its radioprotective action:

1. It is s generally accepted that oxidative stress eventually causes DNA breaks (von Sonntag 2006). Astaxanthin is a powerful antioxidant and can intercept reactive oxygen species induced by ionizing radiation. Antioxidant capacity of astaxanthin probably acts the part in reducing "immediate" DNA damage (we can see that as decrease in TM on 0 hour after irradiation), but hardly affects DNA damage after 48 hours of lymphocyte cultivation and, thereby, it unlikely has a significant influence on ChA formation.
2. In the present study we demonstrated that astaxanthin treatment decreases levels of unstable cytogenetic markers of radiation exposure and relative levels of DNA damage against the background of increasing apoptotic rate. Our data is conformed by previous studies that showed astaxanthin is able to induce apoptosis in cancer cells of many sources of origin (Song et al. 2011, Kavitha et al. 2013, Li et al. 2015, Liu 2016). So, radioprotective action of astaxanthin may be explained rather by its apoptogenic properties (stimulates elimination of "moderately" damaged cells) than by its potential genoprotector capacity (defenses DNA from damage or activates processes of DNA repair).

Conclusion

This study indicated that astaxanthin in chosen concentration had no mutagenic activity and demonstrated evident radioprotective properties. We assume that astaxanthin acts rather as an apoptogen and stimulates elimination of damaged cells than as a veritable genoprotector. The findings in our current research provide important clues toward the possibility of the astaxanthin usage as prophylactic and therapeutic drug under condition of high radiation risk.

References

Afanasieva K, Zazhytska M, Sivolob A. (2010) Kinetics of comet formation in single-cell gel electrophoresis: loops and fragments. *Electrophoresis*. 31:512-519.

Afanasieva K, Chopei M, Zazhytska M, Vikhreva M, Sivolob A. (2013) DNA loop domain organization as revealed by single - cell gel electrophoresis. *Biochim Biophys Acta.* 1833:3237-3244.

Aguilera A, García-Muse T. (2013) Causes of genome instability. *Annu Rev Genet.* 47:1-32.

Ambati RR, Phang SM, Ravi S, Aswathanarayana RG. (2014) Astaxanthin: sources, extraction, stability, biological activities and its commercial applications - a review. *Mar Drugs*.12:128-152.

Ahnström G., Erixon K. (1981) Measurement of strand breaks by alkaline denaturation and hydroxyapatite chromatography, in *DNA Repair. A Laboratory Manual of Research Procedures*, eds Friedberg E. C., Hanawalt P. C., editors. New York, NY: Marcel Dekker. pp. 403–418.

Azqueta A, Collins AR. (2012) Carotenoids and DNA damage. *Mutat Res.* 733:4-13.

Belloni P, Meschini R, Czene S, Harms-Ringdahl M, Palitti F. (2005). Studies on radiation-induced apoptosis in G_0 human lymphocytes. *Int J Radiat Biol.* 81:587-599.

Berti AP, Düsman E, Mariucci RG, Lopes NB, Vicentini VE. (2014). Anti mutagenic and radioprotective activities of beta-carotene against the biological effects of iodine-131 radiopharmaceutical in Wistar rats. *Genet Mol Res*. 13:2248-2258.

Gyori BM, Venkatachalam G, Thiagarajan PS, Hsu D, Clement M. (2014) OpenComet: An automated tool for comet assay image analysis. *Redox Biology*. 2:457–65.

Heijink AM, Krajewska M, van Vugt MA. (2013) The DNA damage response during mitosis. *Mutat Res*. 750:45-55.

Hsu TC, Wu X, Trizna Z. (1996) Mutagen sensitivity in human. A comparison between two nomenclature systems for recording chromatid breaks. *Cancer Genet Cytogenet*, 87:127-132.

Kavitha K, KowshikJ, Kishore TK, Baba AB, Nagini S. (2013) Astaxanthin inhibits NF-kappaB and Wnt/beta-catenin signaling pathways via inactivation of Erk/MAPK and PI3K/Akt to induce intrinsic apoptosis in a hamster model of oral cancer. *Biochim. Biophys. Acta*. 1830:4433–4444.

Kim B, Farruggia C, Ku CS, Pham TX, Yang Y, Bae M, Wegner CJ, Farrell NJ, Harness E, Park YK, Koo SI, Lee JY. (2016) Astaxanthin inhibits inflammation and fibrosis in the liver and adipose tissue of mouse models of diet-induced obesity and nonalcoholic steatohepatitis. *J NutrBiochem*. 43:27-35.

Kim JH, Park JJ, Lee BJ, Joo MK, Chun HJ, Lee SW, Bak YT. (2016) Astaxanthin inhibits proliferation of human gastric cancer cell lines by interrupting cell cycle progression. *Gut Liver*. 10(3):369-374.

Krinsky N. 1994. The biological properties of carotenoids. *Pure Appl Chem*. 66:1003-1006.

Kurinnyi DA, Rushkovsky SR, Demchenko OM, and MA. Pilinska. (2018) Peculiarities of Modification by Astaxanthin of Radiation-Induced Damages in the Genome of Human Blood Lymphocytes Exposed in vitro on Different Stages of the Mitotic Cycle. *Cytology and Genetics*. 52(1):40–45.

Li J, Dai W, Xia Y, Chen K, Li S, Liu T, Zhang R, Wang J, Lu W, Zhou Y, Yin Q, Abudumijiti H, Chen R, Zheng Y, Wang F, Lu J, Zhou Y,

Guo C. (2015) Astaxanthin inhibits proliferation and induces apoptosis of human hepatocellular carcinoma cells via inhibition of Nf-Kb P65 and Wnt/B-Catenin *in vitro*. *Mar Drugs*. 13:6064-6081.

Liao W, McNut MA, Zhu WXG. (2009) The comet assay: A sensitive method for detecting DNA damage in individual cells. *Methods*. 48: 46-53.

Liu X, Song M, Gao Z, Cai X, Dixon W, Chen X, Cao Y, Xiao H. (2016). Stereoisomers of astaxanthin inhibit human colon cancer cell growth by inducing G2/M cell cycle arrest and apoptosis. *J Agric Food Chem*. 64:7750-7759.

Lin KH, Lin KC, Lu WJ, Thomas PA, Jayakumar T, Sheu JR. (2015) Astaxanthin, a carotenoid, stimulates immune responses by enhancing ifn-γ and il-2 secretion in primary cultured lymphocytes *in vitro* and *ex vivo*. *Int J Mol Sci*. 17:44-54.

Little JB. (2002) Genomic instability and radiation. *J Radiol Prot*. 23: 173-181.

Natarajan AT, Palitti F. (2008). DNA repair and chromosomal alterations. *Mutat Res*. 657:3-7.

Nishida Y, Yamashita E, Miki W. (2007). Quenching activities of common hydrophilic and lipophilic antioxidants against singlet oxygen using chemiluminescence detection system. *Car Sci*. 11:16–20.

Obe G, Durante M. (2010). DNA double strand breaks and chromosomal aberrations. *Cytogenet Genome Res*. 128:8-16.

Olive PL, Durand RE. (2005). Heterogeneity in DNA damage using the comet assay. *Cytometry A*. 66:1-8.

Olive PL, Banáth JP. (2006). The comet assay: a method to measure DNA damage in individual cells. *Natprotoc*. 1:23–29.

Palozza P, Torelli C, Boninsegna A, Simone R, Catalano A, Mele MC, Picci N. (2009). Growth-inhibitory effects of the astaxanthin-rich alga *Haematococcuspluvialis* in human colon cancer cells. *Cancer Lett*;283:108-117.

Rao R, Sindhuja H, Dharmesh S, Sankar K, Sarada R, Ravishankar G. (2013). Effective Inhibition of Skin Cancer, Tyrosinase, and Antioxidative Properties by Astaxanthin and Astaxanthin Esters from the Green Alga Haematococcuspluvialis. *J Agric Food Chem.* 61:3842– 3851.

Roser S, Pool-Zobel BL, Rechkemmer G. (2001). Contribution of apoptosis to responses in the comet assay. *Mutat Res*. 497:169-175.

Rosner B. (2015). *Fundamentals of Biostatistics, Eighth Edition*. - Cengage Learning. - 962 p.

Song XD, Zhang JJ, Wang MR, Liu WB, Gu XB, Lv CJ. (2011). Astaxanthin induces mitochondria-mediated apoptosis in rat hepatocellular carcinoma CBRH-7919 cells. *Biol Pharm Bull*. 34:839-844.

von Sonntag C. (2006). *Free-radical-induced DNA damage and its repair. A chemical perspective.* - Springer-Verlag. –p. 528.

Srinivasan M, Devipriya N, Kalpana KB, Menon VP. (2009). Lycopene: An antioxidant and radioprotector against γ-radiation-induced cellular damages in cultured human lymphocytes. *Toxicology*. 262: 43–49.

Tago Y, Fujii T, Wada J, Kato M, Wei M, Wanibuchi H, Kitano M. (2014). Genotoxicity and subacute toxicity studies of a new astaxanthin-containing *Phaffiarhodozyma*extract. *J Toxicol Sci*. 39:373-382.

Tawn EJ, Holdsworth D. (1991). Mutagen induced chromosome damage in human lymphocytes. In: DE. Rooney and BH Czepulkowski (Eds.). *Human Cytogenetics. Volume II. Malignancy and acquired abnormalities. A practical approach.* Irl press, Oxford, New York, Tokyo. 189-208.

Biographical Sketch

Stanislav R. Rushkovsky

Affiliation:

Associated Professor, September 2006 – to present

Department of General and Medical Genetics, Taras Shevchenko National University of Kyiv, Ukraine

Education: Taras Shevchenko National University of Kyiv, Ukraine
Department Gen. & Mol. Genetics
Post-graduate student November 1994-October 1997
Taras Shevchenko Kyiv University Kyiv, Ukraine
M.Sc. in Genetics 1989-1994

Business Address: Department of General and Molecular Genetics,
Taras Shevchenko National University of Kyiv,
64 Volodymyrska St, Kyiv, 01033 Ukraine

Professional: Human genome instability with the special focus on individual features of Appointments: mutagen sensitivity, chromosomal instability and cancer risk, studying of time-dependent mutagenesis (delayed chromosomal instability) and identifying potential antimutagens for cancer prevention.

In: Ionizing Radiation
Editor: Tamar Reeve

ISBN: 978-1-53613-539-8

Chapter 8

AN OVERVIEW OF *in Silico* METHODS FOR THE PREDICTION OF IONIZING RADIATION RESISTANCE IN BACTERIA

***Manel Zoghlami*[1,2,*], *Sabeur Aridhi*[3], *Mondher Maddouri*[4] and *Engelbert Mephu Nguifo*[1]**
[1]Clermont Auvergne University, LIMOS, Clermont-Ferrand, France
[2]University of Tunis El Manar, LIPAH, Tunis, Tunisia
[3]University of Lorraine, LORIA, Nancy, France
[4]University of Jeddah, College Of Business, Jeddah, Kingdom of Saudi Arabia

Abstract

Ionizing-radiation-resistant bacteria (IRRB) could be used for bioremediation of radioactive wastes and in the therapeutic industry. Limited computational works are available for the prediction of bacterial ionizing radiation resistance (IRR). In this chapter, we present some works that study the causes of the high resistance of IRRB to ionizing radiation. Then we focus on presenting *in silico* approaches that use protein sequences of bacteria in order to predict if an unknown bacterium belongs to IRRB or ionizing-radiation-sensitive bacteria (IRSB). These approaches formulate the problem of predicting bacterial IRR as a multiple instance learning (MIL) problem where bacteria represent the bags and

*Corresponding Author: manel.zoghlami@gmail.com.

primary structure of basal DNA repair proteins of each bacterium represent the instances inside the bags. We also present a formulation of the problem of MIL in sequence data and explain how it could be used to solve the problem of IRR prediction in bacteria. A brief comparison of the presented approaches is provided.

Keywords: bacterial ionizing radiation resistance, multiple instance learning, phenotype prediction

1. Introduction

IRRB could be used for the treatment of mixed radioactive wastes by developing a strain to detoxify both mercury and toluene [1]. These organisms are also being engineered for *in situ* bioremediation of radioactive wastes[2]. In [3], the authors discuss the potential uses of radiation-resistant extremophiles (e.g. micro-organisms with the ability to survive in extreme environmental conditions) in biotechnology and the therapeutic industry. The major challenges of therapeutic development using extremophiles are discussed in [4].

Several *in vitro* and *in silico* works studied the causes of the high resistance of IRRB to ionizing radiation to determine peculiar features in their genomes and improve the treatment of radioactive wastes [5]. However, limited computational works are provided for the prediction of bacterial IRR [6][7]. In this work, we present some works that study the causes of the high resistance of IRRB to ionizing radiation. Then we focus on machine learning approaches that use protein sequences of bacteria to predict whether a bacterium belongs to IRRB or IRSB. These approaches formulate the problem of predicting bacterial IRR as an MIL problem where bacteria represent bags and repair proteins of each bacterium represent instances. MIL methods are a variation of machine learning methods that can be used to solve problems in which the labels are assigned to bags, i.e., a set of instances, rather than individual instances.

The remainder of this chapter is organized as follows. In Section 2, we present some computational works that aim to analyze IRRB in order to find out the causes of their resistance to ionizing radiation. In Section 3, we provide a formalization of the problem of MIL and explain how it could be used to solve the IRR prediction problem. Then we present some existing *in silico* approaches for IRR prediction in bacteria. Section 4 provides a brief comparison of the presented approaches. Concluding points make the body of Section 5.

2. Computational Works for Analyzing IRRB

Several *in vitro* and *in silico* works are proposed in order to study the causes of the high resistance of IRRB to ionizing radiation. In fact, determining the characteristics of IRRB could help to predict if a bacterium belongs to IRRB or IRSB. In this chapter, we present some of these works and show the pipeline of existent bioinformatics programs and the statistical methods they use.

In [6], the authors analyze four known genome sequences of IRRB in order to find out the role of positive Darwinian selection in the evolution of IRR and the tolerance of desiccation. The used pipeline contains three steps:
Step 1. Finding orthologous relationships using MultiParanoid [8].
Step 2. Aligning the sequences for each ortholog set using CLUSTAL W [9].
Step 3. Testing for positive selection using the DnaSP program [10] that requires an aligned set of orthologous sequences.
This work concludes that all basal DNA repair genes in IRRB are subject to positive selection unlike many of their orthologs in IRSB.

In [11], the authors make a comparative analysis of codon and amino acid usage patterns. The study uses 19 genomes of the phylum *Deinococcus-Thermus* and follows a six-step pipeline.

Step 1. The authors make correspondence analysis on relative synonymous codon usage and amino acid usage using the program CodonW [12]. Then they use SPSS software [13] to make correlation and variance analysis.
Step 2. In order to find out the variation in amino acid usage between radiation-sensitive and radiation-resistant genome of the studied bacteria, the STATISTICA software [14] is used and a cluster analysis on amino acid usage is performed.
Step 3. To find out a cluster of orthologous proteins, the authors use the CMG-Biotools workbench [15] and a BLAST matrix [16].
Step 4. Synonymous and non-synonymous substitution patterns in previously selected orthologous proteins are then estimated. The PAML software [17] is used in this step.
Step 5. Several indices are calculated in order to find out the factors influencing codon and amino acid usage including aromaticity, average hydrophobicity, isoelectric point and instability index.
Step 6. Finally, a COG [18] functional classification is performed in order to compare the proteins. The IMG/M system [19] is used in this step as a support

for comparative analysis of metagenomes.

This study reports significant differences in synonymous codon usage bias and amino acid usage patterns between the radiation-sensitive and radiation-resistant genomes of the studied dataset.

In a recent work [20], the authors identify general patterns of microbial responses to multiple stressors in radioactive environments. They analyze three datasets including a set of bacteria isolated from soil contaminated by nuclear waste at the Hanford site (USA) [21]. The other two datasets are out of the scope of this chapter since they concern fungi and yeasts. In order to develop a filter procedure for identifying important predictor variables, machine learning and information theoretic approaches are used. The pipeline contains three main steps.

Step 1. The authors start by testing the hypothesis which supposes that the collected contaminated soil samples are statistically dependent (samples separated by 1 meter of soil depth would contain the same bacterial taxa than samples separated by 10 meters). To do so, they use Mantel tests [22] for spatial autocorrelation using the ade4 R package [23].
Step 2. In order to find out the variables which are the least important for describing the data, authors use two machine learning prediction methods: random forests [24] and random generalized linear modeling [25]. The variables that are identified as unimportant by both machine learning methods are most likely to be noise variables and are not used in the rest of the study.
Step 3. All variables are ranked in order of importance and filtered based on information theoretic approaches [26] [27]. A new model is created using the retained variables.

3. Multiple Instance Learning Approaches for Bacterial IRR Prediction

As far as we know, only three *in silico* approaches were proposed in order to predict the bacterial IRR based on a set of protein sequences. These approaches, named ABClass, ABSim and MIL naive approach [28], are three machine learning algorithms that require as input a set of proteins sequences of each bacterium and provide as output a label response: IRRB if the query bacterium is radioresistant and IRSB if not. Since these approaches are based on a MIL formalisation, we present in this section the MIL problem formulation and we highlight some well known MIL algorithms. Then, we provide a description of the above cited approaches.

3.1. Problem Formulation

We denote Σ an *alphabet* defined as a finite set of characters or symbols. A *symbolic sequence* is defined as an ordered list of symbols [29]. The primary structure of a protein is a symbolic sequence since it is described using symbols (amino acids). Let DB be a learning database that contains a set of n labeled bags $DB = \{(B_i, Y_i), i = 1, 2 \ldots, n\}$ where $Y_i = \{-1, 1\}$ is the label of the bag B_i. Instances in B_i are sequences and are denoted by B_{ij}. Formally $B_i = \{B_{ij}, j = 1, 2 \ldots, m\}$, where m is the total number of instances in this bag. We note that according to the problem investigated in this chapter, there is a relationship between instances of different bags since there are orthologous proteins in the different bags. This relation is denoted *the across bag sequences relation* in [28]. The goal is to learn a multiple instance classifier from DB. Given a query bag $Q = \{Q_k, k = 1, 2 \ldots, q\}$, where q is the total number of instances in Q, the classifier should use sequential data in this bag and in each bag of DB in order to predict the label of Q.

3.2. Multiple Instance Learning: An Overview

In a traditional setting of supervised learning task, the training set is composed of feature vectors (instances), where each feature vector has a label. In MIL task, we learn a classifier based on a training set of bags, where each bag contains multiple feature vectors and it is the bag that carries a label [30]. We do not know the labels of the instances inside the bags. The task of MIL investigated in this chapter is to learn a classifier from the training set that correctly predicts unseen bags.

Several MIL algorithms have been proposed. A review of MIL approaches with a comparative study could be found in [30] and [31]. A common assumption in MIL field is that a positive bag contains at least one positive instance, while in a negative bag all of the instances are negative. This assumption is called *the standard multiple instance assumption*. Several MIL algorithms adopt this asymmetrical assumption including MI-SVM [32] and Diverse Density [33]. MI-SVM is an adaptation of support vector machines (SVM) to the

MIL problem. The key point of Diverse Density algorithm is to find a concept point in the feature space that is close to at least one instance from every positive bag and meanwhile far away from instances in negative bags. Other methods adopt the so called *collective assumption* which states that all instances in a bag contribute to define its label. This assumption could be suitable to the problem investigated in this chapter since we do not know which set of proteins helps us to classify a bacterium as IRRB. Some algorithms make a positive instance selection. In a recent work [34], the authors present the MILKDE algorithm which identifies the most representative instances in each positive bag based on a likelihood computation.

According to [30], the MIL methods could be categorized according to how the information existent in the MI data is exploited. Three categories could be defined depending on the adopted paradigm:

- Methods following the instance-space paradigm assume that the discriminative information is present at the instance-level. We consider the characteristics of individual instances in the learning process without looking at more global characteristics of the whole bag. The simplest algorithm in this category is the SIL algorithm [35] which trains a standard supervised classifier then simply uses the sum as aggregation rule to obtain the bag-level classifier.

- Methods following the bag-space paradigm treat each bag as a whole entity. A global bag-level information is used to make the discriminative decision instead of aggregating instance-level decisions. A commonly used approach is to define a distance function in order to compare bags. In [36], the authors present an extension of the classical KNN algorithm [37] called the Citation-kNN algorithm. It classifies a bag based on the labels of its neighbors (references and citers).

- Methods following the embedded-space paradigm map each bag to a feature vector which summarizes the relevant information about the whole bag. The bag-level information is extracted explicitly through the definition of a mapping function, while it is done implicitly in the bag-space paradigm. In [38], the authors propose an MIL algorithm that computes the dissimilarities of a bag to other bags in the training set and uses these dissimilarities as a feature representation.

3.3. A Naive MIL Approach for IRR Prediction

In [28], the authors present a naive MIL approach for sequence data with across bag relationships. Since the bacterial IRR prediction problem could be formalized as an MIL problem and there is a relationship between protein sequences across bags (orthologous proteins), then we can apply the naive approach in our case. This approach contains two steps: a preprocessing step and a learning step.

Step 1. The preprocessing step transforms the set of sequences into an attribute-value matrix where each row corresponds to a sequence and each column corresponds to an attribute. When we deal with sequence data, the most used technique to transform data into an attribute-value format is to extract motifs that serve as attributes. We note that finding a uniform description of all instances using a set of motifs is not always an easy task. Since the naive approach takes into account the across bag relationships between instances, the preprocessing step extracts motifs from each set of related instances. The union of these extracted motifs is then used as an attribute set to construct the descriptive matrix. The presence or the absence of a motif in a sequence is respectively denoted by 1 or 0. It is worthwhile to mention that only a subset of the used attributes is representative for each processed sequence. Therefore, we may have a big sparse matrix when trying to present the whole sequence data using an attribute value format.
Step 2. The second step consists in applying an existing MIL classifier.

3.4. *ABClass*: Across Bag Sequences Classification Approach

In order to avoid the use of one large vector of features to describe sequence data, the ABClass approach [28] that takes into account the across bag relationships between instances is proposed. It contains the following steps.

Step 1. A preprocessing step identifies orthologous protein sequences. Ideally, each protein has an orthologous sequence in each bag. This is defined as an across bag dependency. This relationship between sequences of different bags will be used in the learning step. The learning dataset is divided into sets of orthologous sequences. We note that a protein may not have any ortholog in a bag.
Step 2. A set of motifs is extracted from each set of orthologous sequences. In the experimental tests, DMS [39] is used as a motif extraction method. DMS

allows building motifs that can discriminate a family of sequences from other ones. It first identifies motifs in the protein sequences. Then it filters them in order to keep only the discriminative and minimal ones.
Step 3. Extracted motifs are used to encode instances in order to create a discriminative model. Each set of related instances (i.e. orthologous proteins) is represented by its own motifs vector. This reduces the number of attributes that are not representative for the processed protein sequence. The WEKA [40] data mining tool is used in order to apply existing well known classifiers to generate models. The following three classifiers have been used: SMO [41] [42], J48 [43] and Naive Bayes [44].
Step 4. To predict the IRR of an unlabeled bacterium, the extracted motifs are used to represent its instances. Each protein sequence is represented using a vector. Then it is compared with the corresponding model (already generated from its orthologous proteins) to provide a partial prediction result.
Step 5. Finally, an aggregation step is applied on the partial results in order to compute the final prediction result.

3.5. *ABSim*: Across Bag Sequences Similarity Approach

In [28], the authors present ABSim, an algorithm that discriminates bags by measuring the similarity between each instance sequence in the query bag and corresponding related sequences in the different bags of the learning database. ABSim is an extension of a previously proposed algorithm named MIL-Align [45]. The difference between these two algorithms is that ABSim is more general since MIL-Align was originally proposed to deal with biological data while ABSim aims to deal with MIL in sequence data in general e.g. textual data. When applied to the problem of bacterial IRR prediction, ABSim uses the local alignment score as similarity measure to compare protein sequences. The algorithm works as follows:

Step 1. For each protein sequence in the query bag, the algorithm computes the corresponding alignment scores between this protein and its orthologs in other bags.
Step 2. Alignment scores are grouped into a matrix. Each line corresponds to a score vector of a protein against all its orthologs in other bacteria.
Step 3. An aggregation method is applied to the matrix in order to compute the final prediction result. Two aggregation methods named SMS and WAMS are

proposed by the authors.

3.6. Experimental Results of MIL Approaches

The MIL naive approach, ABClass and ABSim are used in [28] in order to resolve the problem of phenotype prediction of bacterial IRR. Bacteria represent the bags and primary structure of basal DNA repair proteins represent the sequences. An unknown bacterium is affiliated to either IRRB or IRSB. The used dataset is described in [45]. It consists of 28 bacteria (14 IRRB and 14 IRSB). Each bacterium/bag contains 25 to 31 instances that correspond to proteins implicated in basal DNA repair in IRRB. Proteins of the bacterium *Deinococcus radiodurans* were downloaded from the UniProt website. Proteomes of other bacteria were downloaded from the NCBI FTP website. ABSim and ABClass tools can be downloaded at the following link: http://homepages.loria.fr/SAridhi/software/MIL/.

It is worthwhile to mention that ABClass is tested under several different settings (classifiers and motif extraction settings) and that the obtained accuracy results depend on the used settings. The over all accuracy results reported in [28] are very close but a slightly better accuracy rates are provided either by ABClass or by ABSim according to the used settings. Using both approaches, two bacteria (*M. radiotolerans* and *B. abortus*) generate the lowest rates of successful prediction compared to the rate of the other bacteria. These results may help to understand some characteristics of the studied data. A probable biological explanation is provided in [45] and notes that this could be explained by the increased rate of sequence evolution in endosymbiotic bacteria [46].

4. Comparison

Table 4.1 provides a short comparison of the previously presented approaches. The three works [6] [20] and [11] are interested in analyzing different characteristics of IRRB. The works in [6] and [11] are based on a pipeline that uses existent bioinformatics programs in order to perform their studies. The study in [7] is based on a filter of machine learning and information theoretic approaches. Although these works are important to understand the causes of the high resistance of IRRB to ionizing radiation, they do not provide a tool to predict if an unknown bacterium belongs to IRRB or IRSB.

Table 4.1. Comparison of the presented approaches on IRR analyses/prediction in bacteria

Approach	Task	Method
Sghaier et al., 2008 [6]	Analysis of IRRB	Pipeline
Banerjee et al., 2014 [11]	Analysis of IRRB	Pipeline
Shuryak and Dadachova, 2016 [20]	Analysis of IRRB	Information theory and machine learning
MIL Naive Approach [28]	IRR Prediction	MIL classifiers and motif extraction
ABClass [28]	IRR Prediction	MIL model and optimized use of motifs
ABSim [28]	IRR Prediction	MIL model and alignment score

The naive MIL approach, ABClass and ABSim provide as output a prediction label for an unknown bacterium (IRRB or IRSB). The naive MIL approach uses standard MIL classifiers after making a preprocessing step which extracted motifs and adapt the data to the required format. This preprocessing step is time consuming and could lead to a huge matrix. ABCLass takes advantage of the across bag relationships between sequences in order to reduce the number of attributes that are not representative for each sequence during the encoding step. This relationship is also used during the learning step when generating partial models for each set of related sequences. ABSim does not use motifs to represent data since no encoding step is needed. The local alignment score is used to perform the prediction. This makes ABSim faster and easier to use than ABClass unless we already have the representative motifs for each set of orthologous proteins or if we think that the extraction of motifs will not be an expensive task (according to the data size, the used motifs extractor and the extraction settings e.g. required motifs length). As mentioned in the previous section, a slightly better accuracy result could be provided either by ABClass or by ABSim according to the used settings.

Conclusion

Prediction of IRR in bacteria is a challenging task. Several works were interested in finding out the causes of the resistance of some bacteria to IRR.

These works generally use a pipeline of existing bioinformatics programs. Other works provide machine learning algorithms in order to predict the bacterial IRR. They are based on an MIL formalization. In this chapter, we studied existing works on prediction of IRR in bacteria and we presented a comparison of the studied approaches. Based on our study, we mention that the used settings in the preprocessing step and the learning step influence the choice of the approach to use.

References

[1] Brim H, McFarlan SC, Fredrickson JK, Minton KW, Zhai M, Wackett LP, et al. Engineering Deinococcus radiodurans for metal remediation in radioactive mixed waste environments. *Nature biotechnology.* 2000;18(1):85–90.

[2] Brim H, Venkateswaran A, Kostandarithes HM, Fredrickson JK, Daly MJ. Engineering Deinococcus geothermalis for bioremediation of high-temperature radioactive waste environments. *Applied and environmental microbiology.* 2003;69(8):4575–4582.

[3] Gabani P, Singh OV. Radiation-resistant extremophiles and their potential in biotechnology and therapeutics. *Applied microbiology and biotechnology.* 2013;97(3):993–1004.

[4] Singh O, Gabani P. Extremophiles: radiation resistance microbial reserves and therapeutic implications. *Journal of applied microbiology.* 2011;110(4):851–861.

[5] Needleman SB, Wunsch CD. A general method applicable to the search for similarities in the amino acid sequence of two proteins. *Journal of molecular biology.* 1970;48(3):443–453.

[6] Sghaier H, Ghedira K, Benkahla A, Barkallah I. Basal DNA repair machinery is subject to positive selection in ionizing-radiation-resistant bacteria. *BMC genomics.* 2008;9(1):297.

[7] Makarova KS, Omelchenko MV, Gaidamakova EK, Matrosova VY, Vasilenko A, Zhai M, et al. Deinococcus geothermalis: the pool of extreme radiation resistance genes shrinks. *PLoS One.* 2007;2(9):e955.

[8] Alexeyenko A, Tamas I, Liu G, Sonnhammer EL. Automatic clustering of orthologs and inparalogs shared by multiple proteomes. *Bioinformatics.* 2006;22(14):e9–e15.

[9] Thompson JD, Higgins DG, Gibson TJ. CLUSTAL W: improving the sensitivity of progressive multiple sequence alignment through sequence weighting, position-specific gap penalties and weight matrix choice. *Nucleic acids research.* 1994;22(22):4673–4680.

[10] Librado P, Rozas J. DnaSP v5: a software for comprehensive analysis of DNA polymorphism data. *Bioinformatics.* 2009;25(11):1451–1452. Available from: http://www.ub.edu/dnasp/.

[11] Banerjee R, Roy A, Mukhopadhyay S. Genomic and proteomic signatures of radiation and thermophilic adaptation in the Deinococcus-Thermus genomes. *International Journal of Pharmacy and Pharmaceutical Sciences.* 2014;6:287–300.

[12] *Peden J. CodonW software*; 1997. Available from: http://codonw.sourceforge.net/.

[13] SPSS Inc. *SPSS software for windows (version 15.0)*; 2007.

[14] StatSoft Inc. *Statistica 6 software*; 2002. Available from: http://www.statsoft.com/Products/STATISTICA-Features.

[15] Vesth T, Lagesen K, Acar Ö, Ussery D. CMG-biotools, a free workbench for basic comparative microbial genomics. *PLoS One.* 2013;8(4):e60120.

[16] Altschul SF, Gish W, Miller W, Myers EW, Lipman DJ. Basic local alignment search tool. *Journal of molecular biology.* 1990;215(3):403–410.

[17] Yang Z. PAML 4: phylogenetic analysis by maximum likelihood. *Molecular biology and evolution.* 2007;24(8):1586–1591.

[18] Tatusov RL, Natale DA, Garkavtsev IV, Tatusova TA, Shankavaram UT, Rao BS, et al. The COG database: new developments in phylogenetic classification of proteins from complete genomes. *Nucleic acids research.* 2001;29(1):22–28.

[19] Markowitz VM, Chen IMA, Chu K, Szeto E, Palaniappan K, Grechkin Y, et al. IMG/M: the integrated metagenome data management and comparative analysis system. *Nucleic acids research.* 2011;40(1):123–129.

[20] Shuryak I, Dadachova E. *Quantitative modeling of microbial population responses to chronic irradiation combined with other stressors.* PloS one. 2016;11(1):e0147696.

[21] Fredrickson JK, Zachara JM, Balkwill DL, Kennedy D, Shu-mei WL, Kostandarithes HM, et al. Geomicrobiology of high-level nuclear waste-contaminated vadose sediments at the Hanford Site, Washington State. *Applied and environmental microbiology.* 2004;70(7):4230–4241.

[22] Mantel N. The detection of disease clustering and a generalized regression approach. *Cancer research.* 1967;27(2 Part 1):209–220.

[23] Dray S, Dufour AB, et al. The ade4 package: implementing the duality diagram for ecologists. *Journal of statistical software.* 2007;22(4):1–20.

[24] Breiman L. Random forests. *Machine learning.* 2001;45(1):5–32.

[25] Song L, Langfelder P, Horvath S. Random generalized linear model: a highly accurate and interpretable ensemble predictor. *BMC bioinformatics.* 2013;14(1):5.

[26] Burnham KP, Anderson DR. Model selection and multimodel inference: a practical information-theoretic approach. *Springer Science & Business Media*; 2003.

[27] Burnham KP, Anderson DR, Huyvaert KP. AIC model selection and multimodel inference in behavioral ecology: some background, observations, and comparisons. *Behavioral Ecology and Sociobiology.* 2011;65(1):23–35.

[28] Zoghlami M, Aridhi S, Sghaier H, Maddouri M, Nguifo EM. *A multiple instance learning approach for sequence data with across bag dependencies.* CoRR. 2016;abs/1602.00163. Available from: http://arxiv.org/abs/1602.00163.

[29] Xing Z, Pei J, Keogh E. A brief survey on sequence classification. *ACM SIGKDD Explorations Newsletter.* 2010;12(1):40–48.

[30] Amores J. Multiple instance classification: Review, taxonomy and comparative study. *Artificial Intelligence*. 2013;201:81–105.

[31] Alpaydın E, Cheplygina V, Loog M, Tax DM. Single-vs. multiple-instance classification. *Pattern Recognition*. 2015;48(9):2831–2838.

[32] Andrews S, Tsochantaridis I, Hofmann T. Support Vector Machines for Multiple-Instance Learning. In: Thrun, Obermayer K, editors. *Advances in Neural Information Processing Systems*. Cambridge, MA: MIT Press; 2003. p. 561–568.

[33] Maron O, Pérez TL. A Framework for Multiple-Instance Learning. In: Jordan MI, Kearns MJ, Solla SA, editors. *Advances in Neural Information Processing Systems*. vol. 10. Cambridge, MA: The MIT Press; 1998. p. 570–576.

[34] Faria AW, Coelho FGF, Silva A, Rocha H, Almeida G, Lemos AP, et al. MILKDE: A new approach for multiple instance learning based on positive instance selection and kernel density estimation. *Engineering Applications of Artificial Intelligence*. 2017;59:196–204.

[35] Bunescu RC, Mooney RJ. Multiple instance learning for sparse positive bags. In: *Proc. 24th international conference on Machine learning*. ACM; 2007. p. 105–112.

[36] Wang J. Solving the multiple-instance problem: A lazy learning approach. In: *Proc. 17th International conference on Machine Learning*. Morgan Kaufmann; 2000. p. 1119–1125.

[37] Shakhnarovich G, Darrell T, Indyk P. Nearest-neighbor methods in learning and vision. *IEEE transactions on neural networks*. 2008;19(2):377.

[38] Cheplygina V, Tax DM, Loog M. Multiple instance learning with bag dissimilarities. *Pattern Recognition*. 2015;48(1):264–275.

[39] Maddouri M, Elloumi M. Encoding of Primary Structures of Biological Macromolecules Within a Data Mining Perspective. *Journal of Computer Science and Technology*. 2004;19(1):78–88.

[40] Hall M, Frank E, Holmes G, Pfahringer B, Reutemann P, Witten IH. The WEKA data mining software: an update. *ACM SIGKDD explorations newsletter*. 2009;11(1):10–18.

[41] Platt J. Fast training of support vector machines using sequential minimal optimization. In: *Advances in kernel methods*. MIT Press; 1999. p. 185–208.

[42] Keerthi SS, Shevade SK, Bhattacharyya C, Murthy KRK. Improvements to Platt's SMO algorithm for SVM classifier design. *Neural computation*. 2001;13(3):637–649.

[43] Quinlan JR. *C4. 5: programs for machine learning*. Morgan Kaufmann Publishers; 1993.

[44] John GH, Langley P. Estimating continuous distributions in Bayesian classifiers. In: *Proc. 11th conference on uncertainty in artificial intelligence*. Morgan Kaufmann Publishers Inc.; 1995. p. 338–345.

[45] Aridhi S, Sghaier H, Zoghlami M, Maddouri M, Nguifo EM. Prediction of Ionizing Radiation Resistance in Bacteria Using a Multiple Instance Learning Model. *Journal of Computational Biology*. 2016;23(1):10–20.

[46] Woolfit M, Bromham L. Increased rates of sequence evolution in endosymbiotic bacteria and fungi with small effective population sizes. *Molecular Biology and Evolution*. 2003;20(9):1545–1555.

INDEX

I

L

M

N

O

P

R